72 Advances in Polymer Science
Fortschritte der Hochpolymeren-Forschung

Epoxy Resins and Composites I

Editor: K. Dušek

With Contributions by
A. Apicella, J. M. Barton, N. S. Eiss,
A. J. Kinloch, J. E. McGrath, R. J. Morgan,
L. Nicolais, C. Tran, G. L. Wilkes,
E. M. Yorkgitis

With 79 Figures and 16 Tables

Springer-Verlag
Berlin Heidelberg New York Tokyo

ISBN-3-540-15546-5 Springer-Verlag Berlin Heidelberg New York Tokyo
ISBN-0-387-15546-5 Springer-Verlag New York Heidelberg Berlin Tokyo

Library of Congress Catalog Card Number 61-642

This work is subject to copyright. All rights are reserved, whether the whole or part of the material is concerned, specifically those of translation, reprinting, re-use of illustrations, broadcasting, reproduction by photocopying machine or similar means, and storage in data banks. Under § 54 of the German Copyright Law where copies are made for other than private use, a fee is payable to "Verwertungsgesellschaft Wort". Munich.

© Springer-Verlag Berlin Heidelberg 1985
Printed in GDR

The use of registered names, trademarks, etc. in this publication does not imply, even in the absence of a specific statement, that such names are exempt from the relevant protective laws and regulations and therefore free for general use.

Typesetting and Offsetprinting: Th. Müntzer, GDR;
Bookbinding: Lüderitz & Bauer, Berlin

2154/3020-543210

Editors

Prof. Henri Benoit, CNRS, Centre de Recherches sur les Macromolecules, 6, rue Boussingault, 67083 Strasbourg Cedex, France

Prof. Hans-Joachim Cantow, Institut für Makromolekulare Chemie der Universität, Stefan-Meier-Str. 31, 7800 Freiburg i. Br., FRG

Prof. Gino Dall'Asta, Via Pusiano 30, 20137 Milano, Italy

Prof. Karel Dušek, Institute of Macromolecular Chemistry, Czechoslovak Academy of Sciences, 16206 Prague 616, ČSSR

Prof. John D. Ferry, Department of Chemistry, The University of Wisconsin, Madison, Wisconsin 53706, U.S.A.

Prof. Hiroshi Fujita, Department of Macromolecular Science, Osaka University, Toyonaka, Osaka, Japan

Prof. Manfred Gordon, Department of Pure Mathematics and Mathematical Statistics, University of Camebridge CB2 1SB, England

Prof. Gisela Henrici-Olivé, Chemical Department, University of California, San Diego, La Jolla, CA 92037, U.S.A.

Prof. Dr. habil. G. Heublein, Sektion Chemie, Friedrich-Schiller-Universität, Humboldtstraße 10, 69 Jena, DDR

Prof. Dr. H. Höcker, Universität Bayreuth, Makromolekulare Chemie I, Universitätsstr. 30, 8580 Bayreuth, FRG

Prof. Hans-Henning-Kausch, Laboratoire de Polymères, Ecole Polytechnique Fédérale de Lausanne, 32, ch. de Bellerive, 1007 Lausanne, CH

Prof. Joseph P. Kennedy, Institute of Polymer Science, The University of Akron, Akron, Ohio 44325, U.S.A.

Prof. Anthony Ledwith, Department of Inorganic, Physical and Industrial Chemistry. University of Liverpool, Liverpool L69 3BX, England

Prof. Seizo Okamura, No. 24, Minamigoshi-Machi Okazaki, Sayko-Ku. Kyoto 606, Japan

Prof. Salvador Olivé, Chemical Department, University of California, San Diego, La Jolla, CA 92037, U.S.A.

Prof. Charles G. Overberger, Department of Chemistry. The University of Michigan, Ann Arbor, Michigan 48104, U.S.A.

Prof. Helmut Ringsdorf, Institut für Organische Chemie, Johannes-Gutenberg-Universität, J.-J.-Becher Weg 18–20, 6500 Mainz, FRG

Prof. Takeo Saegusa, Department of Synthetic Chemistry, Faculty of Engineering, Kyoto University, Kyoto, Japan

Prof. Günter Victor Schulz, Institut für Physikalische Chemie der Universität, 6500 Mainz, FRG

Prof. William P. Slichter, Chemical Physics Research Department, Bell Telephone Laboratories, Murray Hill, New Jersey 07971, U.S.A.

Prof. John K. Stille, Department of Chemistry. Colorado State University, Fort Collins, Colorado 80523, U.S.A.

Editorial

With the publication of Vol. 51 the editors and the publisher would like to take this opportunity to thank authors and readers for their collaboration and their efforts to meet the scientific requirements of this series. We appreciate the concern of our authors for the progress of "Advances in Polymer Science" and we also welcome the advice and critical comments of our readers.

With the publication of Vol. 51 we would also like to refer to a editorial policy: *this series publishes incited, critical review articles of new developments in all areas of polymer science in English (authors may naturally also include workes of their own)*. The responsible editor, that means the editor who has invited the author, discusses the scope of the review with the author on the basis of a tentative outline which the author is asked to provide. The author and editor are responsible for the scientific quality of the contribution.

Manuscripts must be submitted in content language and form satisfactory to Springer-Verlag. Figures and formulas should be reproducible. To meet the convenience of our readers, the publisher will include "volume index" which characterizes the content of the volume.

The editors and the publisher will make all efforts to publish the manuscripts as rapidly as possible, i.e., at the maximum six months after the submission of an accepted paper. Contributions from diverse areas of polymer science must occasionally be united in one volume. In such cases a "volume index" cannot meet all expectations, but will nevertheless provide more information than a mere volume number.

Starting with Vol. 51, each volume will contain a subject index.

Editors   Publisher

Preface

This volume of ADVANCES IN POLYMER SCIENCE contains the first part of a series of critical reviews on selected topics concerning epoxy resins and composites. The last decade has been marked by an intense development of applications of epoxy resins in traditional and newly developing areas such as coatings, adhesives, civil engineering or electronics and high-performance composites. The growing interest in applications and requirements of high quality and performance has provoked a new wave in fundamental research in the area of resin synthesis, curing systems, properties of cured products and methods of their characterization.

The collection of reviews to be published in ADVANCES IN POLYMER SCIENCE is devoted just to these fundamental problems. The epoxy resin-curing agent formulations are typical thermosetting systems of a rather high degree of complexity. Therefore, some of the formation-structure-properties relationships are still of empirical or semiempirical nature. The main objective of this series of articles is to demonstrate the progress in research towards the understanding of these relationships in terms of current theories of macromolecular systems.

Because of the complexity of the problems discussed, the theoretical approaches and interpretation of results presented by various authors and schools may be somewhat different. It may be hoped, however, that a confrontation of ideas may positively contribute to the knowledge about this important class of polymeric materials.

In view of the wide range of this area, it was impossible to publish all contributions in successive volums of ADVANCES IN POLYMER SCIENCE. Part I is published in this Vol. 72; Part II will appear in Vol. 75. Part III and Part IV will follow in the beginning of 1986.

The reader may appreciate receiving a list of all contributions to the series EPOXY RESINS AND COMPOSITES to appear in ADVANCES IN POLYMER SCIENCE:

M. T. Aronhime and J. K. Gillham (Princeton University, Princeton, N.J., USA)
The Time-Temperature-Transformation (TTT) Cure Diagram of Thermosetting Polymeric Systems

A. Apicella and L. Nicolais (University of Naples, Naples, Italy)
Effect of Water on the Properties of Epoxy Matrix and Composites (Part I, Vol. 72)

J. M. Barton (Royal Aircraft Establishment, Farnborough, UK)
The Application of Differential Scanning Calorimetry (DSC) to the Study of Epoxy Resins Curing Reactions (Part I, Vol. 72)

W. Burchard (University of Freiburg, Freiburg i. Br., FRG)
Branching in Epoxy Resins Based on Diglycidyl Ethers of Bisphenol A

L. T. Drzal (Michigan State University, East Lansing, MI, USA)
The Interphase in Epoxy Composites (Part II, Vol. 75)

K. Dušek (Institute of Macromolecular Chemistry, Czechoslovak Academy of Sciences, Prague, Czechoslovakia)
Network Formation in Curing of Epoxy Resins

M. Fedtke (Technical University, Merseburg, GDR)
Elucidation of the Mechanism of Epoxy Curing by Model Reactions

A. Gupta (Jet Propulsion Laboratory, Caltech, Pasadena, CA, USA)
Mechanism and Kinetics of the Cure Process in Tetraglycidylmethane Dianiline-Diaminodiphenyl Sulphone Thermoset System

T. Kamon and H. Furukawa (The Kyoto Municipal Research Institute of Industry, Kyoto, Japan)
Curing Mechanism and Mechanical Properties of Cured Epoxy Resins

J. L. Kardos and M. P. Duduković (Washington University, St. Louis. MO, USA)
Void Growth and Transport During Processing of Thermosetting Matrix Composites

A. J. Kinloch (Imperial College, London, UK)
Mechanics and Mechanisms of Fracture of Thermosetting Epoxy Polymers

E. S. W. Kong (Hewlett-Packard Laboratories, Palo Alto, CA, USA)
Physical Aging in Epoxy Matrices and Composites

J. D. LeMay and F. N. Kelley (University of Akron, Akron, OH, USA)
Structure and Ultimate Properties of Epoxy Resins

F. Lohse, and H. Zweifel (Ciba-Geigy, Basle, Switzerland)
Photocrosslinking of Epoxy Resins

J. A. Manson, R. W. Hertzberg, G. Attalla, D. Shah, J. Hwang and J. Turkanis (Lehigh University, Bethlehem, PA, USA)
Fatigue in Neat and Rubber-Modified Epoxies

E. Mertzel and J. L. Koenig (Case Western Reserve University, Cleveland, OH, USA)

Application of FT-IR and NMR to Epoxy Resins (Part II, Vol. 75)

R. J. Morgan (Lawrence Livermore National Laboratory, Livermore, CA, USA)

Structure-Properties Relations of Epoxies Used as Composite Matrices (Part I, Vol. 72)

E. F. Oleinik (Institute of Chemical Physics, Academy of Sciences of USSR, Moscow, USSR)

Structure and Properties of Epoxy-Aromatic Amine Networks in the Glassy State

B. A. Rozenberg (Institute of Chemical Physics, Academy of Sciences of USSR, Moscow, USSR)

Kinetics, Thermodynamics and Mechanism of Reactions of Epoxy Oligomers with Amines (Part II, Vol. 75)

S. D. Senturia and N. F. Sheppard (Massachusetts Institute of Technology, Cambridge, MA, USA)

Dielectric Analysis of Epoxy Cure

R. G. Schmidt and J. P. Bell (University of Connecticut, Storrs, CT, USA)

Epoxy Adhesion to Metals (Part II, Vol. 75)

E. M. Yorkgitis, N. S. Eiss, Jr., C. Tran, G. L. Wilkes and J. E. McGrath (Virginia Polytechnic Institute, Blacksburg, VA, USA)

Siloxane Modified Epoxy Resins (Part I, Vol. 72)

The editor wishes to express his gratitude to all contributors for their cooperation.

Prague, August 1985

Karel Dušek
Editor

Table of Contents

Structure-Property Relations of Epoxies Used as Composite Matrices
R. J. Morgan 1

Mechanics and Mechanisms of Fracture of Thermosetting Epoxy Polymers
A. J. Kinloch 45

Effect of Water on the Properties of Epoxy Matrix and Composite
A. Apicella, L. Nicolais 69

Siloxane-Modified Epoxy Resins
E. M. Yorkgitis, N. S. Eiss, Jr., C. Tran, G. L. Wilkes, J. E. McGrath 79

The Application of Differential Scanning Calorimetry (DSC) to the Study of Epoxy Resins Curing Reactions
J. M. Barton 111

Author Index Volumes 1–72 155

Subject Index 165

Structure-Property Relations of Epoxies Used as Composite Matrices*

R. J. Morgan
Lawrence Livermore National Laboratory, L-338
University of California, P.O. Box 808,
Livermore, California, 94550 U.S.A.

The structure-deformation/failure process-mechanical property relations of epoxies used as matrices in high performance fibrous composites are presented. Such composites are fabricated either from carbon fiber-epoxy prepregs or by filament winding. The parameters that affect the processing, cure reactions and the resultant chemical and physical structure of the epoxies are discussed. The deformation and failure processes of these glasses are described. The structural parameters that control the deformation and failure processes, the mechanical response and aging of epoxies are addressed and means of improving their processing and performance are described.

1 Introduction	3
2 Processing and Chemical Structure of Epoxies Used in Filament Wound Composites	3
2.1 Introduction	3
2.2 Cure Reactions	4
2.3 Improved Filament Winding Epoxies	4
3 Processing and Chemical Structure of TGDDM-DDS Epoxies Used in Composites Fabricated from Prepregs	6
3.1 Introduction	6
3.2 Starting Materials-Impurities	7
3.3 NMR Characterization of BF_3 : Amine Catalysts	7
3.3.1 Introduction	7
3.3.2 Chemical Composition	8
3.3.3 Thermal Stability	9
3.3.4 Hydrolytic Stability	11
3.3.5 Interaction of BF_3 : $NH_2C_2H_5$ with DDS and TGDDM	11
3.3.6 Catalyst Composition in Prepregs	12
3.3.7 Catalytic Species and Activity	14
4 DSC Studies of the Cure Reactions	15
4.1 Introduction	15
4.2 Prepreg Mixtures and Their Components	15

---

* This work performed under the auspices of the U. S. Department of Energy by the Lawrence Livermore National Laboratory under Contract W-7405-Eng-48

	4.3 Environmental Sensitivity of $BF_3:NH_2C_2H_5$ Catalyst	18
	4.4 Commercial Prepregs	18

5 FTIR Studies of the Cure Reactions 18
    5.1 Introduction 18
    5.2 TGDDM Epoxide Homopolymerization 19
    5.3 TGDDM-DDS Cure Reactions 22
    5.4 Rates and Chemistry of Cure Reactions 28
    5.5 Prepreg Processing Viscosity 30

6 Physical Structure 31
    6.1 Introduction 31
    6.2 Macroscopic Inhomogeneities 31
    6.3 Free Volume 31
    6.4 Network Structure 32

7 Deformation and Failure Modes 35

8 Structural Parameters that Control Mechanical Properties 38

9 Service Environment Aging 39

10 References 40

1 Introduction

The increasing use of high-performance fibrous composites in critical structural applications has led to a need to predict the lifetimes of these materials in service environments. To predict the durability of a composite in service environment requires a basic understanding of (1) the microscopic deformation and failure processes of the composite; (2) the significance of the fiber, epoxy matrix and fiber-matrix interfacial region in composite performance; and (3) the relations between the structure, deformation and failure processes and mechanical response of the fiber, epoxy matrix and their interface and how such relations are modified by environmental factors.

In this paper we review our studies on the structure-property relations of epoxy matrices used in high-performance fibrous composites that are fabricated either from C fiber-epoxy prepregs or by filament winding. We consider the parameters that affect the processing, cure reactions and resultant chemical and physical structure of these crosslinked glasses. The structural parameters that control the deformation and failure modes, mechanical response and service environment aging of these epoxies are addressed. We, also, discuss epoxy systems that exhibit superior processing, thermal and mechanical properties relative to those presently utilized as composite matrices.

2 Processing and Chemical Structure of Epoxies Used in Filament Wound Composites

2.1 Introduction

Epoxy resins utilized in forming filament-wound composites must possess low viscosities (η's) and long gel-times at 23 °C. To minimize unreacted starting materials in the finally cured composite requires the chemical cure reactions of the epoxy system must be simple. Furthermore, the number of chemical starting components in the resin must be small to minimize mixing problems that would result in variable thermal and mechanical properties. The toxicity of the resin chemical starting materials must be low. Also, the epoxy system must attain full cure at relatively low post-cure temperatures, <150 °C, to minimize the development of fabrication strains in the composite.

Amine-cured diglycidyl ether of bisphenol-A (DGEBA) epoxies are the principal matrices used in filament wound composites. Pure DGEBA, DER332 (Dow)[1] epoxide cured with an aliphatic polyethertriamine, Jeffamine T403 (Jefferson) is a commonly used epoxy for filament wound Kevlar 49 composites. The chemical structures of the amine and epoxide monomers are shown in Fig. 1. We have studied the structure-property relations of the DGEBA-T403 epoxy in some detail [1].

---

1 Reference to a company or product name does not imply approval or recommendation of the product by the University of California or the U.S. Department of Energy to the exclusion of others that may be suitable

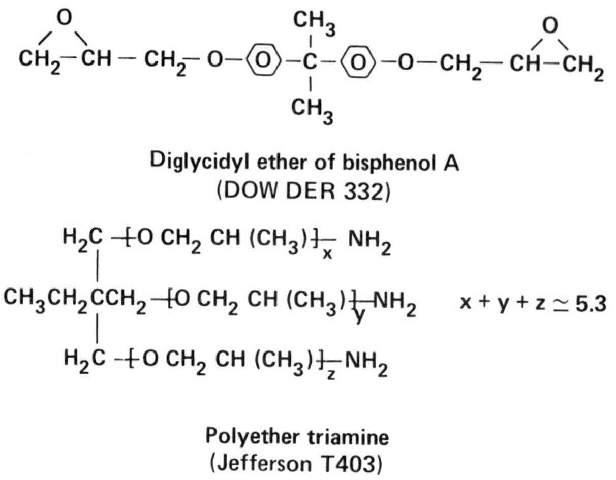

$$CH_2-CH-CH_2-O-\langle O \rangle-\underset{CH_3}{\overset{CH_3}{C}}-\langle O \rangle-O-CH_2-CH-CH_2$$
$$\overset{O}{\diagdown\diagup}\overset{O}{\diagdown\diagup}$$

**Diglycidyl ether of bisphenol A
(DOW DER 332)**

$$H_2C-[O\ CH_2\ CH\ (CH_3)]_x-NH_2$$
$$|$$
$$CH_3CH_2CCH_2-[O\ CH_2\ CH\ (CH_3)]_y-NH_2 \qquad x+y+z \simeq 5.3$$
$$|$$
$$H_2C-[O\ CH_2\ CH\ (CH_3)]_z-NH_2$$

**Polyether triamine
(Jefferson T403)**

Fig. 1. Chemical structure of DGEBA epoxide and T403 polyethertriamine curing agent

2.2 Cure Reactions

The amine-cured DGEBA epoxies utilized as matrices for filament wound composites generally form exclusively from epoxide-amine addition reactions (1).

$$R_1CH-CH_2 + R_2NH_2 \rightarrow R_1CH(OH)\ CH_2NHR_2 \qquad (1)$$
$$\diagdown\diagup$$
$$O$$

The nature of the cure reactions in these epoxies can be confirmed by monitoring the epoxide consumption via near infra-red spectroscopy for a series of epoxide-amine mixtures containing a range of amine contents. A plot of % epoxide consumption vs. amine concentration for DGEBA-T403 epoxies is illustrated in Fig. 2. This plot confirms that the DGEBA-T403 epoxy system forms exclusively from epoxide-amine addition reactions, because (i) 100% epoxide consumption is attained at the stoichiometric amine concentration associated with exclusive epoxide-amine addition cure reactions and (ii) extrapolation of this plot to zero amine content indicates there is no epoxide consumption i.e. there are no epoxide homopolymerization reactions.

Characterization of the epoxy cure reactions ensures that a composite can be fabricated and the epoxy is fully cured, assuming that the epoxide and amine starting components are initially homogeneously mixed.

2.3 Improved Filament Winding Epoxies

The toughness and mechanical performance of a filament wound composite component is enhanced by crack deflection mechanisms and/or molecular flow occurring in the

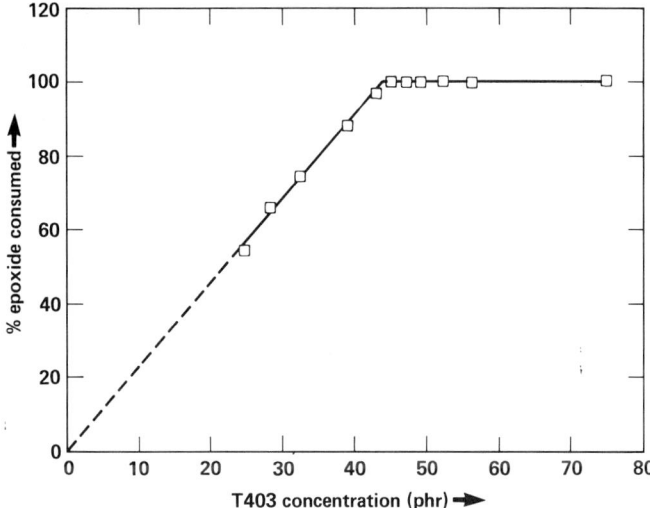

Fig. 2. Epoxide consumption vs. T403 amine concentration for DGEBA-T403 epoxies cured at 85 °C, 24 h

epoxy matrix. The inhibition of crack propagation through a bundle of fibers in a composite can occur by deflection of the crack parallel to the fiber axis by either propagation along the fiber matrix interface and/or through the fiber itself. A poor fiber-matrix interfacial bond and/or microscopic fiber failure by splitting will both enhance these crack deflection toughening mechanisms. The composite performance can also be enhanced under load by molecular flow occurring in the epoxy matrix. Molecular flow is enhanced as T_g is approached because the glassy-state free volume is increased. However, the epoxy matrix cannot be too soft, otherwise the composite will readily buckle in compression. In the case of filament-wound Kevlar 49-epoxy composites the poor fiber-interfacial strength, the microscopic splitting of the fibers and matrix ductility all enhance composite mechanical performance [2, 3]. However, for C-fiber-epoxy filament wound composites the fiber-matrix interface is generally stronger than for Kevlar 49 composites and the C-fiber fails without longitudinal fiber splitting. Hence, for C-fiber epoxy composites the matrix is the principal component that affects composite toughness, and this matrix must be tough through a wide temperature range and possess a $T_g > 120$ °C. This requires that epoxy matrices for filament-wound C-fiber structures are (i) processible at 23 °C, (ii) fully reacted upon post-curing < 150 °C, (iii) simple chemical and physical systems with limited toxicity and (iv) tough from 23–125 °C.

To attain the requirements of an epoxy matrix utilized in filament-wound C-fiber-epoxy composites we have considered the characteristics required of the amine curing agent molecule. To ensure long gel times at 23 °C requires that the primary amine-epoxide (P.A.-E) reaction rate is considerably greater than the rate of the secondary amine-epoxide (S. A.-E) reaction, and that the S.A. reaction does not occur at low temperatures. Furthermore, to attain low 23 °C η's and low post-cure temperatures (to achieve full cure) the amine molecule must be relatively flexible. It was determined that 2,5-dimethyl-2,5-hexanediamine (DMHDA),

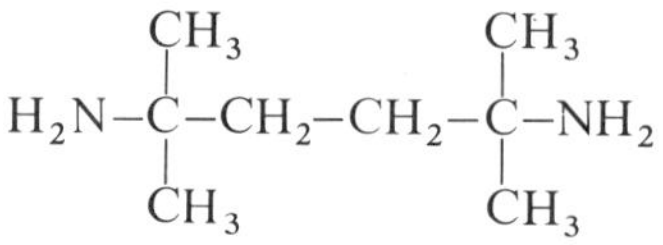

$$H_2N-\underset{CH_3}{\overset{CH_3}{C}}-CH_2-CH_2-\underset{CH_3}{\overset{CH_3}{C}}-NH_2$$

which is a relatively flexible molecule with a sterically crowded C atom adjacent to the NH_2 group fulfills these requirements.[4,5] The DGEBA-DMHDA epoxy system exhibits long gel-times and low η's at 23 °C. The P.A.-E reaction is $\sim 50 \times$ faster than the S.A.-E reaction. Full cure can be attained within 3 h at 100 °C and 90% epoxide consumption can occur even at 60 °C after 120 h. This epoxy system is ductile in the 23–125 °C temperature range and exhibits a T_g of 143 °C.

3 Processing and Chemical Structure of TGDDM-DDS Epoxies Used in Composites Fabricated from Prepregs

3.1 Introduction

Diaminodiphenyl sulfone (DDS) cured tetraglycidyl-4,4′-diaminodiphenylmethane (TGDDM) epoxies are the most common composite matrices utilized in high performance fibrous composites prepared from prepregs. The structures of the unreacted TGDDM epoxide and DDS monomers are illustrated in Fig. 3. The TGDDM epoxide monomer is a liquid at 23 °C, whereas the DDS monomer is a crystalline

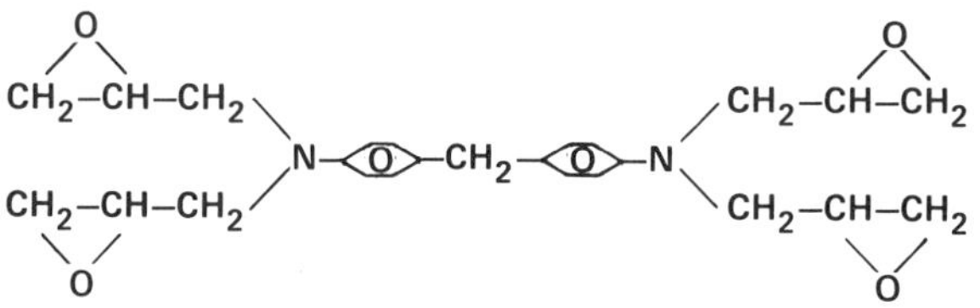

Tetraglycidyl 4, 4′ diaminodiphenyl methane epoxy TGDDM
(liquid at 23°C)

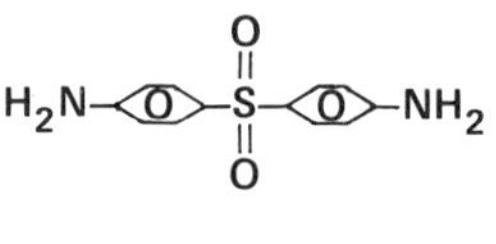

4, 4′ diaminodiphenyl sulfone DDS
(crystalline solid, mp 162°C)

Fig. 3. The structure of TGDDM and DDS monomers

powder with a m.p. of ~165 °C. The commercially available prepreg resins such as Narmco 5208, Fiberite 934 and Hercules 3501, all primarily consist of the TGDDM-DDS epoxy; the latter two systems also contain boron trifluoride catalysts [6-8].

To manufacture reproducible C-fiber-TGDDM-DDS epoxy composites with well-defined lifetimes in service environment requires a knowledge of the parameters that affect composite processing conditions and the resultant structure of the epoxy within the composite. The cure reactions directly control the composite processing and the final epoxy network structure and mechanical response. Hence, it is important to understand the cure reactions and the variables that affect such reactions.

In this Sect. we describe the starting material impurities and their effect on the processing and cure reactions of TGDDM-DDS epoxies. The cure reactions are characterized by differential scanning calorimetry (DSC) and Fourier transform infrared spectroscopy (FTIR) studies. The BF_3:amine catalysts used to accelerate the cure of TGDDM-DDS epoxies are characterized by nuclear magnetic resonance (NMR) spectroscopy studies.

3.2 Starting Materials-Impurities

The commercially available TGDDM, DDS and BF_3:amine components all contain impurities, some of which may act as catalysts towards the cure reactions.

FTIR studies indicate that commercial TGDDM (MY720, Ciba Geigy) contains 15–20% less epoxide groups than the pure tetrafunctional TGDDM epoxide molecule [9-11]. Liquid chromatography studies indicate that the missing

$-CH_2-\overset{\overset{\displaystyle O}{\diagup\diagdown}}{CH}-CH_2$ epoxide groups are replaced by (i) $-CH_2-CH(OH)-CH_2(OH)$, (ii) $-CH_2-CH(OH)-CH_2Cl$, (iii) $-H$, (iv) $-CH_3$ and (v) $-CH=CH-CH_2Cl$ groups with the α-glycol being the predominant impurity species [12-14]. In addition higher oligomers and homopolymer species may also be present. Pearce et al. [13], Hagnauer and Pearce [14] and Scola [15] report that the epoxide groups in TGDDM can hydrolyze to the α-glycol species and that such species catalyze homopolymerization reactions.

Commercial DDS also contains a small percentage of a crystalline impurity as indicated by a DSC endotherm at 77 °C whose heat of fusion is 3% of the total heat of fusion associated with the pure DDS m.pt at ~165 °C [16].

The impurities in the BF_3:amine catalysts are highly variable and are discussed in detail in the following Sect. 3.3.

3.3 NMR Characterization of BF_3:Amine Catalysts

3.3.1 Introduction

The cure reactions, the viscosity-time-temperature profile, the processing conditions, the resultant epoxy chemical and physical structure, and the mechanical response of a C-fiber/TGDDM-DDS cured epoxy composite are modified by the presence of a BF_3-amine complex catalyst within the prepreg. These factors also will be modified by the distribution of the catalyst within the prepreg, its chemical composition, and any modification of its structure and activity as a result of exposure to or interactions with heat or both, moisture, and the epoxide and amine components within the prepreg.

The two most common BF_3:amine catalysts used commercially to cure epoxies are boron trifluoride monoethylamine, $BF_3:NH_2C_2H_5$, and boron trifluoride piperidine, $BF_3:NHC_5H_{10}$, complexes. Such complexes are latent catalysts at room temperature but enhance epoxide group reactivity at higher temperatures.

In this Sect. we discuss ^1H, ^{19}F and ^{11}B NMR studies of $BF_3:NH_2C_2H_5$ and $BF_3:NHC_5H_{10}$ complexes, with principal emphasis on the former. We present the chemical composition of commercial BF_3:amine complexes, their thermal stability in the solid state and in solution, the effect of moisture and heat upon their composition, the nature of their interaction with the epoxide and amine components utilized in TGDDM-DDS commercial prepregs, the composition of BF_3:amine complexes in commercial prepregs, their thermal stability in the prepregs, and the chemical structure of the predominant catalytic species of the cure reactions of the prepreg.

3.3.2 Chemical Composition

The ^1H NMR spectrum of $BF_3:NH_2C_2H_5$ should exhibit peaks in three separate regions, namely, the CH_3 region at highest field, the CH_2 region at an intermediate field, and the NH_2 region at lowest field. The theoretical peak intensity distribution should be 3:2:2 for the CH_3, CH_2, and NH_2 regions, respectively. The ^1H NMR spectra for the four $BF_3:NH_2C_2H_5$ samples are illustrated in Fig. 4. Two components were observed in the spectra. The major component spectra consisted of a CH_3 triplet (1.090 ppm), a CH_2 quartet (2.612 ppm), and an NH_2 signal (6.156 ppm). The inten-

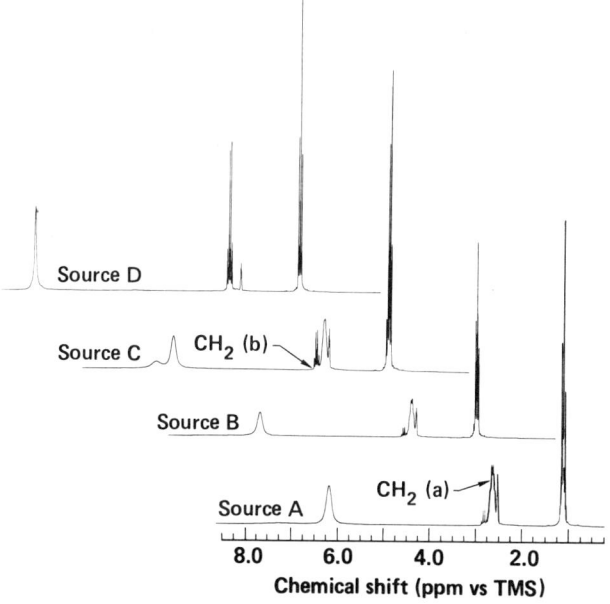

Fig. 4a and b. ^1H NMR spectra of commercial $BF_3:NH_2C_2H_5$ samples (A = Harshaw, B = Alfa, C = K. and K., D = Pfaltz and Bauer); (a) and (b) illustrate two different CH_2 components

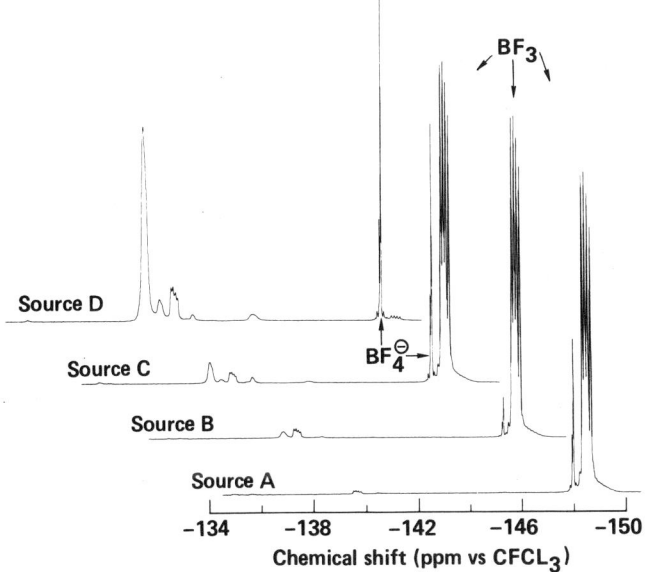

Fig. 5. ^{19}F NMR spectra of commercial $BF_3:NH_2C_2H_5$ samples (A = Harshaw, B = Alfa, C = K. and K., D = Pfaltz and Bauer)

sity ratios were 3:2:2, respectively, and this spectrum was assigned to the $BF_3:NH_2C_2H_5$. A second component was evident in the proton NMR spectra, characterized again by a CH_3 triplet (1.121 ppm), a CH_2 quartet (2.803 ppm), and an NH/OH peak from labile protons (7.27 to 7.63 ppm). Intensity ratios were 3:2:3. This spectrum is attributed to (BF_4^- or BF_3OH^-) $NH_3^+C_2H_5$ species. The proton intensity measurements indicate that BF_4^- is the dominant anion.

The ^{19}F NMR spectra of the commercial $BF_3:NH_2C_2H_5$ samples are illustrated in Fig. 5. The major components identified in the ^{19}F spectra were $BF_3:NH_2C_2H_5$, BF_4^-, and $BF_3(OH)^-$ species and an unidentified highly reactive BF_3 species with an NMR peak in the region of $BF_3(OH)^-$. The fluorine species observed in the commercial samples are illustrated in Table 1, in which several other observed ^{19}F NMR peaks are combined under the "Miscellaneous" heading.

The chemical composition of a $BF_3:NHC_5H_{10}$ sample was also investigated by ^{19}F NMR. Three fluorine-containing environments were found with fluorine distributed as follows: $BF_3:NHC_5H_{10}$ (87.3%), $BF_4^-NH_2^+C_5H_{10}$ (11.5%), and $BF_3(OH)^-NH_2^+C_5H_{10}$ (0.2%).

3.3.3 Thermal Stability

Solid $BF_3:NH_2C_2H_5$ samples that were annealed at 85, 115, or 139 °C for 1 h and then subsequently dissolved in dimethyl sulfoxide (DMSO) exhibited no significant dissociation as detected by ^{1}H. These data are consistent with observations by Harris and Temin [18] that BF_3:amine complexes do not dissociate irreversibly to gaseous BF_3 and amine products. (The $BF_3:NH_2C_2H_5$ was observed to melt near 85 °C during these studies.)

Table 1. Fluorine species in $BF_3:NH_2C_2H_5$ from ^{19}F NMR

Source	Percent Total Fluorine as				
	$BF_3:NH_2C_2H_5$	BF_4^-	$BF_3(OH)^-$	Reactive BF_3	Miscellaneous
Alfa	92.5	2.5	2.9	2.0	0.1
Pfaltz and Bauer	1.9	15.3	12.4	57.3	13.1
Harshaw	89.7	8.9	0.9	—	0.5
K. and K.	78.1	11.1	3.1	4.8	2.9

However, ^{19}F NMR studies indicate that a small amount of the $BF_3:NH_2C_2H_5$ may slowly convert to BF_4^- and to another species which we do not detect in the ^{19}F spectrum. There is an apparent loss of fluorine as illustrated in Table 2. The small percentage of $BF_3(OH)^-$ species present in the unannealed sample disappears after a 1-h annealing at 85 °C, presumably as a result of either reaction with species at the glass sample tube surface or formation of

Table 2. Effect of thermal annealing on the fluorine species in $BF_3:NH_2C_2H_5$ from ^{19}F NMR

Annealing Conditions	Percent Total Original Fluorine			
	Fluorine Loss	$BF_3(OH)^-$	BF_4^-	$BF_3:NH_2C_2H_5$
Original (unannealed solid)	0	6.1	1.6	92.3
1 h at 85 °C	8.1	0	11.8	80.1
1 h at 115 °C	2.7	0	13.8	83.5
1 h at 140 °C	9.7	0	13.7	76.6

species which undergo chemical exchange at an intermediate rate on the NMR time scale.

If $BF_3:NH_2C_2H_5$ is heated directly in DMSO, the conversion to BF_4^- and the percentage fluorine loss is considerably greater than if $BF_3:NH_2C_2H_5$ is heated in the absence of the solvent, as illustrated in Table 3. The conversion of the $BF_3:NH_2C_2H_5$ to BF_4^- species with associated fluorine loss could perferentially

Table 3. Effect of heating on the fluorine species in $BF_3:NH_2C_2H_5$ DMSO solution from ^{19}F NMR

Heat Conditions	Percent Total Original Fluorine as			
	Fluorine Loss	$BF_3(OH)^-$	BF_4^-	$BF_3:NH_2C_2H_5$
Unheated Solution	0	6.1	1.6	92.3
1 h at 85 °C	5.1	0	13.1	81.8
1 h at 115 °C	12.9	0	34.7	52.4
1 h at 140 °C	27.3	0	60.2	12.5

occur at the glass sample tube surface. Hence, such reactions would be accelerated in solution because the mobility of the $BF_3:NH_2C_2H_5$ is enhanced. However, we cannot rule out the possibility of a reaction between the $BF_3:NH_2C_2H_5$ and the DMSO solvent that enhances conversion to BF_4^- species.

$BF_3:NHC_5H_{10}$ exhibits similar thermal stability trends as $BF_3:NH_2C_2H_5$.

3.3.4 Hydrolytic Stability

^{19}F NMR studies of $BF_3:NH_2C_2H_5$/DMSO solutions indicated that little change occurred in the amount of $BF_3:NH_2C_2H_5$, BF_4^-, and $BF_3(OH)^-$ species in the presence of added H_2O over a period of 4 days at 23 °C.

However, if a large (ten-fold) excess of H_2O was added to the $BF_3:NH_2C_2H_5$/DMSO solution and the temperatures raised to 85 °C for 1 h, the amount of fluorine present as $BF_3:NH_2C_2H_5$ decreased by 50% while that present as $BF_3(OH)^-NH_3^+C_2H_5$ increased from 5 to 40%. The more stable BF_4^- species were unaffected.

The $BF_3NHC_5H_{10}$ species in DMSO behaved similarly to the corresponding $BF_3:NH_2C_2H_5$ species upon exposure to H_2O at 23 and 85 °C.

3.3.5 Interaction of $BF_3:NH_2C_2H_5$ with DDS and TGDDM

^1H and ^{19}F NMR were used to study the interaction of the individual resin components of the C-fiber/TGDDM-DDS prepregs with $BF_3:NH_2C_2H_5$.

It was determined that the ^1H NMR spectrum of DDS in DMSO does not change when $BF_3:NH_2C_2H_5$ is added at ambient temperature. $BF_3:NH_2C_2H_5$/DDS/DMSO solutions were then monitored after heating for 1 h at 85, 115, or 139 °C. The broad CH_2 multiple associated with $BF_3:NH_2C_2H_5$ in the proton spectra decreased with increasing temperature exposure, ultimately resulting in a sharp CH_2 quartet which is associated with either $BF_4^-NH_3^+C_2H_5$ or $BF_3(OH)^-NH_3^+C_2H_5$ species.

^{19}F NMR studies also indicate that significant changes in the BF_3 species occur upon heating $BF_3:NH_2C_2H_5$/DDS/DMSO solutions. The degree of conversion to the BF_4^- salt, for the same thermal exposure, is similar to that observed in $BF_3:NH_2C_2H_5$/DMSO solution (Table 3) in the absence of DDS. Hence, we have no direct evidence that DDS competes with $C_2H_5NH_2$ for BF_3 molecules or that DDS enhances BF_4^- salt formation upon heating in DMSO solution.

The ^1H NMR spectra of $BF_3:NH_2C_2H_5$/TGDDM/DMSO solutions were investigated as a function of thermal exposure. Heating the solution for 1 h at 85 °C did not produce changes in the ^1H spectra. However, exposures to 115 or 139 °C for 1 h did produce significant spectral changes. The unmodified TGDDM ^1H NMR spectrum contains two doublets centered at 7.022 and 6.763 ppm which are associated with the two types of aromatic protons. The five different chemical proton environments associated with the

$$N-CH_2-\overset{\overset{\displaystyle O}{\triangle}}{CH}-CH_2$$

group result in the series of peaks in the 2.500- to 3.500-ppm region. We have not attempted to assign peaks in this region to specific proton environments. The essential disappearance of the TGDDM aromatic proton doublets and the modification of the spectral region associated with the TGDDM aliphatic protons after heating for 1 h at 115 and 139 °C confirms that $BF_3:NH_2C_2H_5$ reacts extensively with the TGDDM epoxide.

^{19}F NMR studies, illustrated in Table 4, indicate that the TGDDM epoxide enhances BF_4^- salt formation in DMSO solution. For example, after exposure to 115 °C for 1 h, all the $BF_3:NH_2C_2H_5$ species have disappeared in the presence of TGDDM. However, in DMSO solution in the absence of TGDDM, 50% of the total fluorine species are still in the form of $BF_3:NH_2C_2H_5$ after 1 h exposure at 115 °C (Table 3).

Table 4. Effect of heating on the fluorine species in $BF_3:NH_2C_2H_5$/TGDDM/DMSO solution from ^{19}F NMR

Heat Conditions	Percent Total Original Fluorine as			
	Fluorine Loss	$BF_3(OH)^-$	BF_4^-	$BF_3:NH_2C_2H_5$
Unheated solution	0	6.1	1.6	92.3
1 h at 85 °C	19.4	0	17.8	62.8
1 h at 115 °C	23.0	0	77.0	0
1 h at 140 °C	20.6	0	79.4	0

3.3.6 Catalyst Composition in Prepregs

The catalyst composition in Fiberite 934 and Hercules 3501 prepregs was investigated by ^{19}F NMR. The epoxy resin in these commercial C-fiber/TGDDM-DDS prepregs was dissolved in DMSO.

The fluorine species observed in five different lots of Fiberite 934 were identified and tabulated in Table 5. A typical spectrum is illustrated in Fig. 6. The catalyst in

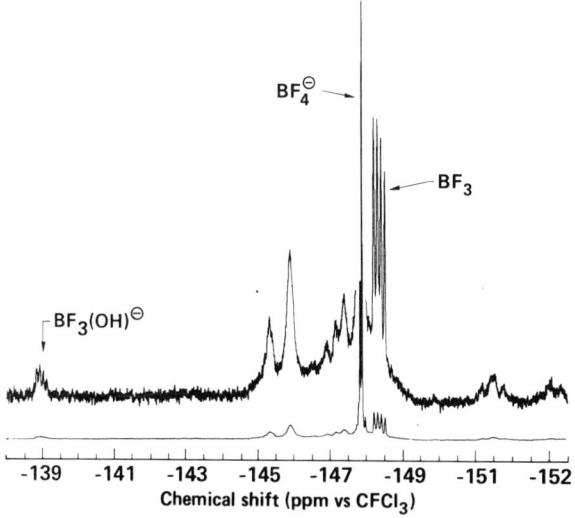

Fig. 6. ^{19}F NMR spectrum of DMSO-extracted components from Fiberite 934 prepreg

this prepreg was identified as $BF_3:NH_2C_2H_5$ by ^{19}F NMR. There is considerable variation from lot to lot in the $BF_4^-NH_3^+C_2H_5$ (14.6 to 60.0%) and the $BF_3:NH_2C_2H_5$ (7.2 to 48.6%) species.

Table 5. Fluorine species in Fiberite 934 lots

Lots	$BF_3(OH)^-$ $NH_3^+C_2H_5$	$BF_4^-NH_3^+C_2H_5$	$BF_3:NH_2C_2H_5$	Epoxide-BF_3 Products
C2-709	3.8	38.8	25.2	32.2
C3-218	4.6	60.0	7.2	28.2
C3-389	0.5	14.6	48.6	36.3
C3-397	1.5	44.4	22.4	31.7
C3-546	1.4	49.1	18.2	31.3

We associate the various additional fluorine peaks observed in the NMR spectra with principally the products of epoxide-active fluorine species reactions. The total fluorine in the form of these products is relatively constant from prepreg lot to lot (28.2 to 36.3%).

The epoxy resin of a Hercules 3501 sample was dissolved in DMSO and investigated by ^{19}F NMR. The ^{19}F peaks associated with $BF_3:NH_2C_2H_5$ were absent and the 1:1:1:1 quartet associated with $BF_3:NHC_5H_{10}$ was found at an ^{19}F chemical shift value of -155.17 ppm. The fluorine distribution among species found in this sample was $BF_3(OH)^-NH_2^+C_5H_{10}$ (2.8%), $BF_4^-NH_2^+C_5H_{10}$ (9.9%), $BF_3:NHC_5H_{10}$ (78.3%), and epoxide-fluorine products (9.0%).

The Fiberite 934 prepreg was exposed to a series of temperature-time profiles and the soluble epoxy resin portion dissolved in DMSO and studied by ^{19}F NMR. The ratios of the intensity of the peak associated with $BF_3:NH_2C_2H_5$ (I_{BF_3}) to that intensity of the $BF_4^-NH_3^+C_2H_5$ ($I_{BF_4^-}$) species as a function of exposure conditions are tabulated in Table 6. With increasing annealing temperature, the $BF_3:NH_2C_2H_5$ species concentration decreases relative to the $BF_4^-NH_3^+C_2H_5$ species. The presence of steam slows the relative disappearance of the $BF_3:NH_2C_2H_5$ species.

Table 6. The effect of temperature, time, and H_2O on the relative extractable $BF_3:NH_2C_2H_5$ and $BF_4^-NH_3^+C_2H_5$ species in Fiberite 934

Exposure Conditions	$I_{BF_3}/I_{BF_4^-}$
Ambient	0.718; 0.748
1 h at 50 °C	0.539
5 h at 50 °C	0.554
1 h at 75 °C	0.494
1 h at 100 °C	0.146
1 h at 100 °C + steam	0.326

3.3.7 Catalytic Species and Activity

Our NMR studies indicate that $BF_3:NH_2C_2H_5$ is slowly converted to $BF_4^- NH_3^+ C_2H_5$ salt with corresponding loss of fluorine upon heating the solid catalyst. This conversion to the salt is accelerated in DMSO solution and further accelerated in TGDDM/DMSO solutions with an associated 20 to 30% loss of fluorine upon near-complete conversion to the salt. We will now consider the catalytic mechanism and activity of $BF_3:NH_2C_2H_5$ toward the TGDDM-DDS cure reaction in the light of our NMR observations.

The $BF_3:NH_2C_2H_5$ can react directly with an epoxide resulting the formation of a monoboroester and HF

$$BF_3-NH_2C_2H_5 + R-\overset{O}{\overset{\diagup\diagdown}{CH-CH_2}} \rightarrow$$
$$\downarrow$$
$$R-\underset{\underset{OBF_2}{|}}{CH}-CH_2-NH-C_2H_5 + HF \qquad (2)$$

The HF generated then reacts with another $BF_3:NH_2C_2H_5$ to form the salt

$$BF_3-NH_2C_2H_5 + HF \rightarrow BF_4^- NH_3^+ C_2H_5 \qquad (3)$$

The HF can also react with the components of the prepreg resulting in a variety of carbon-fluorine containing species. The formation of the BF_4^- salt requires two $BF_3:NH_2C_2H_5$ molecules to be in close proximity. Hence, the formation of the BF_4^- salt will intimately depend on the dispersion of the small quantity (0.4 wt.-%) of the $BF_3:NH_2C_2H_5$ catalyst within the prepreg and, therefore, could be highly variable. The monoboroester can act as a catalyst for the cure reactions [19] but is susceptible to deactivation by hydrolysis.

All BF_3 and BF_3-prepregs species, with the exception of the $BF_4^- NH_3^+ C_2H_5$ salt are susceptible to transformation to less active or nonactive catalytic species. Furthermore, our NMR results indicate that the $BF_4^- NH_3^+ C_2H_5$ salt does not irreversibly chemically react with the prepreg components.

For each epoxide group to be catalyzed by a BF_3 species in a prepreg containing 0.4 wt.-% $BF_3:NH_2C_2H_5$ catalyst requires each catalytic species to act as a catalyst to 200 epoxide groups. This means each BF_3 catalytic species has to be chemically stable and mobile. Hence, we suggest that the $BF_4^- NH_3^+ C_2H_5$ salt is the predominant catalytic species for the prepreg cure reactions, with the more active BF_3 species becoming deactivated or immobilized during the early stages of cure. Harris and Temin [18] have reported that BF_3-amine complexes and their corresponding BF_4^- salts cure epoxides in the same temperature range and cure times. This observation is consistent with the BF_4^- salt being the predominant catalytic species and the BF_3-amine complex converting to the BF_4^- salt in the presence of epoxide groups.

4 DSC Studies of the Cure Reactions

4.1 Introduction

In C-fiber-TGDDM-DDS prepregs that do not contain a BF_3:amine catalyst one exotherm peak associated with the cure reactions has been observed by DSC [20-22]. However, prepregs that contain BF_3:amine catalyst have been reported to exhibit 2 or 3 additional DSC peaks at lower temperatures, which have been associated with the catalyzed cure reactions [20, 23].

In this Sect. we report systematic DSC studies of (i) the constituents of boron trifluoride monoethylamine ($BF_3:NH_2C_2H_5$) catalyzed TGDDM-DDS epoxies and their mixtures; (ii) the nature of the catalyzed cure reactions and (iii) the environmental sensitivity of the $BF_3:NH_2C_2H_5$ catalyst. DSC studies are also reported on the cure reaction characteristics and environmental sensitivity of commercial prepregs that contain BF_3:amine catalysts.

4.2 Prepreg Mixtures and their Components

In Figs. 7a and b DSC plots are compared for a standard TGDDM (64 wt.-%)/DDS (25 wt.-%)/DGOP (11 wt.-%) commercial prepreg mixture containing 0 and 0.4% of the $BF_3:NH_2C_2H_5$ catalyst respectively, (DGOP is a diglycidyl orthophthalate epoxide, Gly-Cel-A-100, Celanese). In the absence of the $BF_3:NH_2C_2H_5$ catalyst (Fig. 7a) a large exotherm occurs at 240 °C ($\Delta H = 470$ j/g) with a smaller exotherm at 125 °C ($\Delta H = 30$ j/g), whereas upon addition of the catalyst (Fig. 7b) 4 peaks designated α, β, γ and δ are present at 240 °C ($\Delta H = 210$ j/g); 200 °C ($\Delta H = 190$ j/g); 160 °C ($\Delta H = 70$ j/g) and 125 °C ($\Delta H = 25$ j/g) respectively.

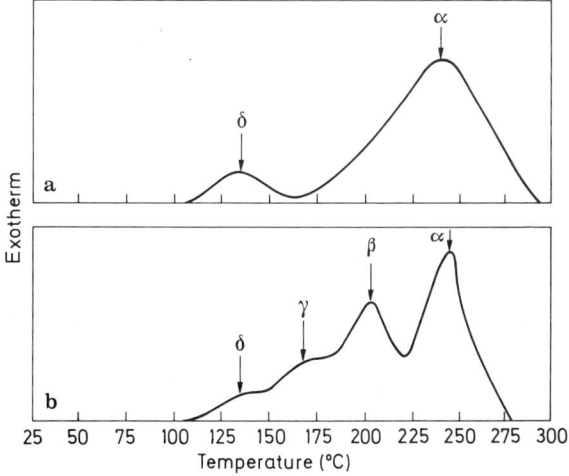

Fig. 7a and b. DSC plots for TGDDM (64 wt.-%)/DDS (25 wt.-%)/DGOP (11 wt.-%) epoxy prepreg mixture with (a) 0.0 wt.-% and (b) 0.4 wt.-% $BF_3:NH_2C_2H_5$ catalyst

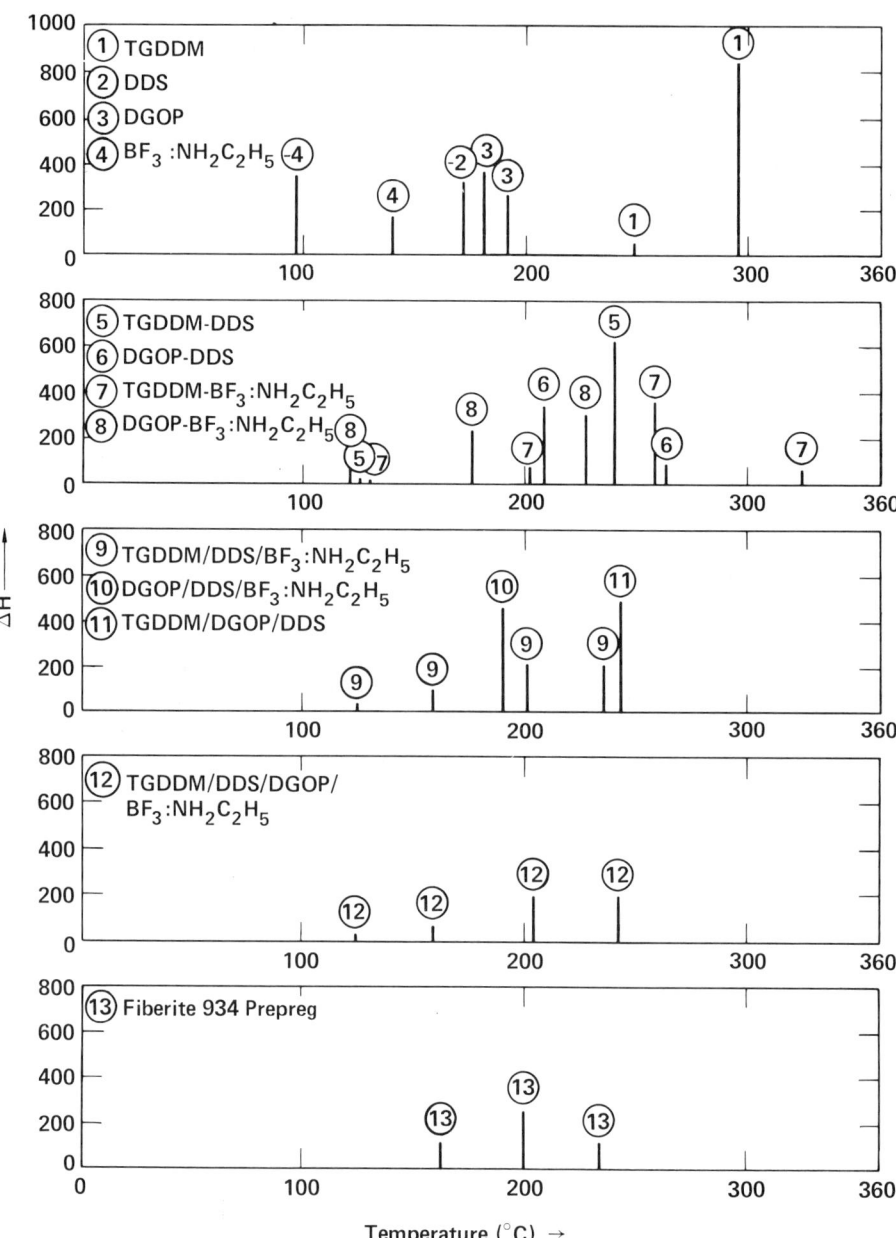

Fig. 8. DSC ΔH peak intensities in j/g versus the temperature associated with maximum peak intensity for 1, 2, 3 and 4 component mixtures of TGDDM/DDS/DGOP/$BF_3:NH_2C_2H_5$ epoxies and commercial Fiberite 934 prepreg

In an effort to characterize the chemical reactions associated with each peak, DSC runs were performed on each of the constituents of the epoxy prepreg and systematic 2, 3 and 4 constituent mixtures. (The constituent mixtures were formulated at the same relative compositions that were present in a standard epoxy prepreg mixture). The ΔH values and temperatures associated with maximum peak intensity for each DSC peak for each component and epoxy prepreg mixture were determined and are plotted in Fig. 8. The DSC peaks are represented by lines in the ΔH-temperature plots in Fig. 8, with the magnitude of each line representing the ΔH value associated with each peak, and its position on the temperature scale representing the temperature of maximum peak intensity. All line magnitudes represent exotherm peaks with the exception of the DDS and $BF_3:NH_2C_2H_5$ endotherm melting points which are designated by negative symbols (-2) and (-4), respectively. The five ΔH-temperature plots in Fig. 8 represent, from the top to the bottom of the Figure, 1, 2, 3 and 4 component prepreg mixtures and Fiberite 934 respectively. From this systematic study illustrated in Fig. 8 we conclude the α peak at 240 °C is associated with the non-catalyzed cure reactions and the β, γ and δ peaks with $BF_3:NH_2C_2H_5$ catalyst-epoxy prepreg constituent reactions.

The ΔH values associated with the α, β and γ peaks are plotted as a function of $BF_3:NH_2C_2H_5$ catalyst concentration in Fig. 9. The α peak intensity associated with non-catalyzed cure reactions progressively decreases with increasing $BF_3:NH_2C_2H_5$ concentration and approaches zero near 2 wt.-% $BF_3:NH_2C_2H_5$. The γ peak intensity progressively increases with $BF_3:NH_2C_2H_5$ concentration, whereas the β peak intensity attains a maximum intensity at ~ 0.4 wt.-% $BF_3:NH_2C_2H_5$ and then decreases with higher concentrations of $BF_3:NH_2C_2H_5$. The small δ peak intensity does not appear to be modified by increasing $BF_3:NH_2C_2H_5$ concentration.

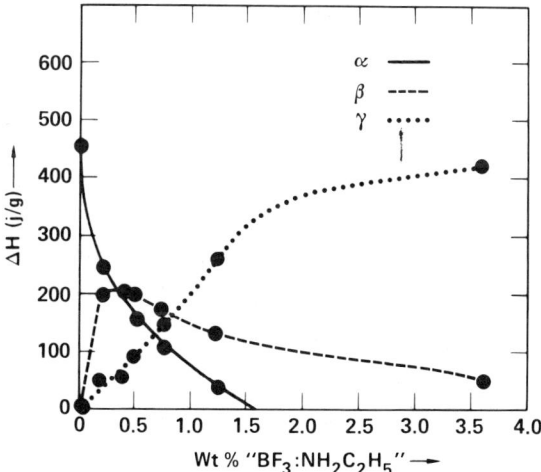

Fig. 9. ΔH values for the DSC α, β and γ peaks in TGDDM (64 wt.-%)/DDS (25 wt.-%)/DGOP (11 wt.-%) epoxies as a function of $BF_3:NH_2C_2H_5$ concentration

From these DSC studies together with our NMR observations from Section 3.3 we conclude (i) the δ peak is associated with $BF_3:NH_2C_2H_5$ catalyzed DDS-TGDDM impurity reactions; (ii) the γ peak is associated $BF_3:NH_2C_2H_5$ and monofluroborate catalyses of the cure reactions and (iii) the β peak is associated $BF_4^-NH_3^+C_2H_5$ cationic catalyses of the cure reactions.

4.3 Environmental Sensitivity of $BF_3:NH_2C_2H_5$ Catalyst

There are reports in the literature of H_2O deactivating the catalytic activity of $BF_3:NH_2C_2H_5$ [24,25]. Our NMR studies [17] indicate that $BF_3:NH_2C_2H_5$ can hydrolyze to form the hydroxy fluoroborate salt, particularly at more extreme conditions of 85 °C. DSC peak intensities indicate that when the $BF_3(OH)^-NH_3^+C_2H_5$ salt is substituted for $BF_3:NH_2C_2H_5$ in a TGDDM/DDS prepreg mix the cure reactions are not modified. This suggests that the $BF_3(OH)^-NH_3^+C_2H_5$ dehydrates back to $BF_3:NH_2C_2H_5$ during the early stages of cure. However, exposure of the Fiberite 934 prepreg to 85 °C for 1 h at 100% RH does produce a significant shift in the intensities of the DSC peaks with the γ peak intensity decreasing by ~50% and the α peak intensity increasing by 50%. We suggest that during the environmental exposure conditions we may have leached a portion of the $BF_3:NH_2C_2H_5$ catalyst out of the prepreg, as we have observed the catalyst is readily soluble in H_2O.

4.4 Commercial Prepregs

We investigated the variability in catalyst activity within a Fiberite 934 lot and also between different lots by monitoring the magnitude of the ΔH values associated with the α, β and γ DSC peaks. The ΔH values associated with each peak varied by only ±5% for samples within close proximity of each other (<10 cms apart) within a prepreg lot. However, the variability in ΔH values was considerably greater (up to ±30%) for samples investigated from (i) the same prepreg lot that were widely separated (>20 cm apart) and (ii) between different prepreg lots.

From DSC ΔH peak intensities we ascertained that on the average Fiberite 934 prepreg cure reactions occur (i) 25% by non-catalyzed reactions (α peak), (ii) 50% $BF_4^-NH_3^+C_2H_5$ catalyzed reactions (β peak) and (iii) 25% by $BF_3:NH_2C_2H_5$ catalyzed reactions (γ peak). From similar DSC studies on the Hercules 3501 prepreg we concluded the cure reactions occur (i) 75% by $BF_4^-NH_2C_5H_{10}$ catalyzed reactions and (ii) 25% by $BF_3:NHC_5H_{10}$ catalyzed reactions. In both commercial prepregs the δ peak is absent, because the BF_3:amine catalyzed DDS-TGDDM impurity reactions have already occurred during the epoxy mixing and C-fiber-prepreg processing conditions.

5 FTIR Studies of the Cure Reactions

5.1 Introduction

To consume all epoxide groups in the TGDDM-DDS epoxy system exclusively by P.A.-E and S.A.-E addition reactions would require 37 wt.-% DDS. However, commercial TGDDM-DDS systems contain only 20–25 wt.-% DDS. Thus, other reactions as well as epoxide-amine addition reactions must occur to consume all the epoxide groups in these systems.

In this Section we report systematic FTIR studies of the cure reactions of TGDDM-DDS epoxies. We report our cure and degradation reaction studies of TGDDM-DDS epoxies from 100 to 300 °C as a function of cure time, and DDS and $BF_3:NH_2C_2H_5$ concentration [9–11, 26, 27].

5.2 TGDDM Epoxide Homopolymerization

The homopolymerization reactions of impure TGDDM (MY720) in the presence and absence of a $BF_3:NH_2C_2H_5$ catalyst and, also, pure TGDDM were monitored by FTIR as a function of cure temperature from 177 to 300 °C. The intensities of the epoxide, hydroxyl, ether and carbonyl bands at 906, 3500, 1120 and 1720 cm^{-1} respectively were determined from spectral differences and are plotted as a function in cure conditions in Figs. 10, 11, 12 and 13 respectively. The 906, 1120 and 1720 cm^{-1} band intensities were normalized to the 805 cm^{-1} band and the 3500 cm^{-1} to the 1615 cm^{-1} band. The 805 and 1615 cm^{-1} bands are associated with the phenyl group which is assumed to chemically unmodified during the homopolymerization reactions.

Epoxide consumption, Fig. 10, primarily occurs in the 175–250 °C range with the rate of epoxide consumption being MY720-$BF_3:NH_2C_2H_5$ > MY720 > pure TGDDM. In agreement with previous observations [13–15], the impurities in MY720 enhance epoxide consumption relative to pure TGDDM.

Hydroxyl, ether and carbonyl band intensities (Figs. 11, 12, and 13) simultaneously increase in the same temperature range as the epoxide band intensity decreases (Fig. 10). The carbonyl band appears to be directly correlated to epoxide consumption rather than general oxidation reactions because the carbonyl band intensity does not increase once the epoxide groups are all consumed. The intensities of the ether,

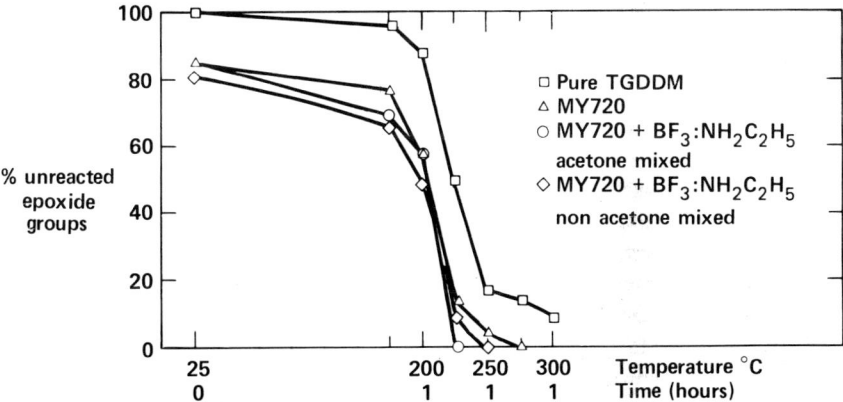

Fig. 10. % of unreacted epoxide groups vs. cure conditions for pure TGDDM (□), MY720 (△), MY720 + 0.4 wt.-% $BF_3:NH_2C_2H_5$-non-acetone mixed (◇), MY720 + 0.4 wt.-% $BF_3:NH_2C_2H_5$-acetone mixed (○)

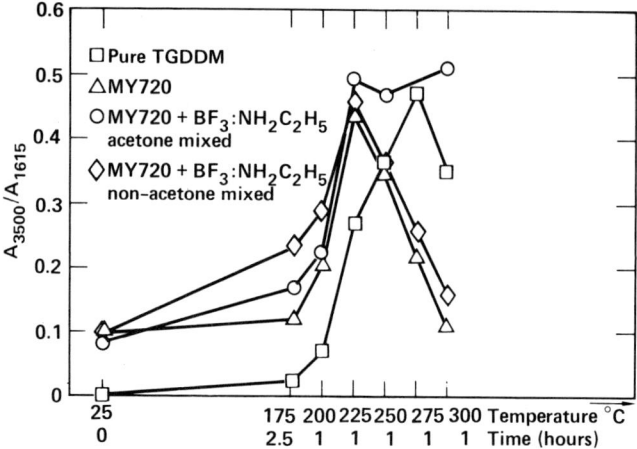

Fig. 11. Hydroxyl group IR intensity (A_{3500}/A_{1615}) vs. cure conditions for pure TGDDM (□), MY720 (△), MY720 + 0.4 wt.-% $BF_3:NH_2C_2H_5$-non-acetone mixed (◇), MY720 + 0.4 wt.-% $BF_3:NH_2C_2H_5$-acetone mixed (○).

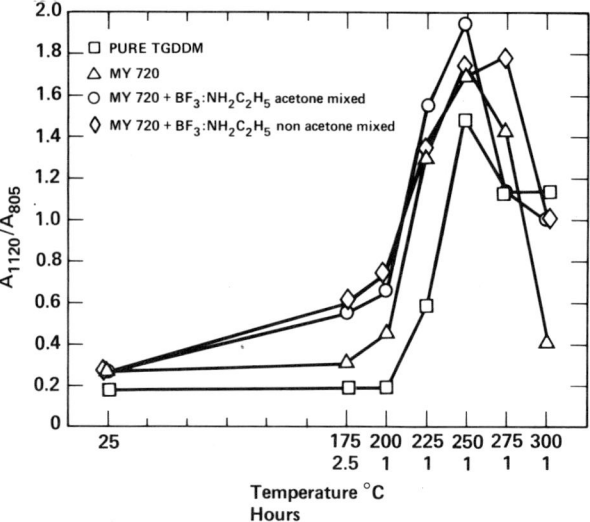

Fig. 12. Ether group IR intensity (A_{1120}/A_{805}) vs. cure conditions for pure TGDDM (□), MY720 $BF_3:NH_2C_2H_5$-non-acetone mixed (◇), MY720 + 0.4 wt.-% $BF_3:NH_2C_2H_5$-acetone mixed (○)

hydroxyl and carbonyl bands decrease with increasing temperature in the 225–300 °C range as a result of network degradation.

In the 177–300 °C temperature range studied, epoxide isomerization, oxidation and homopolymerization can occur followed by complex degradation reactions. There have been numerous studies on the homopolymerization of epoxides including the effects of catalysts, alcohols, cure temperature and epoxide-amine ratio on the

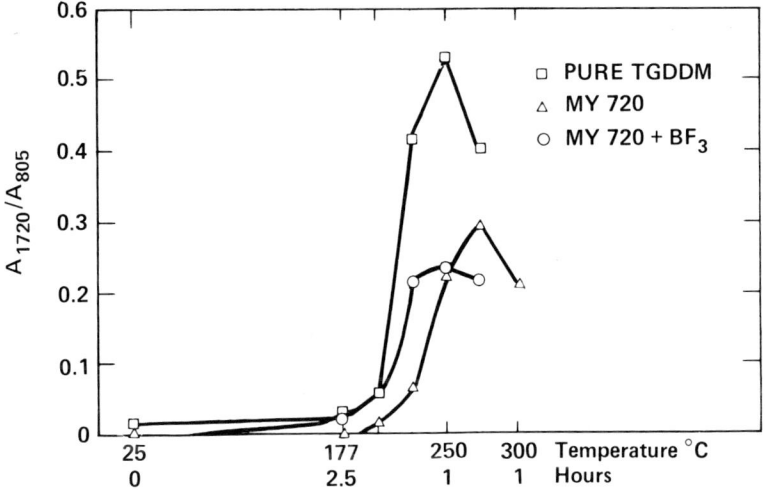

Fig. 13. Carbonyl group IR intensity (A_{1720}/A_{805}) vs. cure conditions for pure TGDDM (□), MY720 (△), MY720 + 0.4 wt.-% $BF_3:NH_2C_2H_5$-acetone and non-acetone mixed (○)

chain extension reactions [13, 14, 20, 28–49]. The appearance of hydroxyl, ether and unsaturation have been reported during homopolymerization of epoxides [30, 37, 40].

There is considerable evidence in the literature that chain extension reactions can occur between epoxide and hydroxyl groups and that this reaction (4) is enhanced in the presence of the tertiary amines [28, 29, 32, 36, 40, 42–48, 50]. In the case of TGDDM

$$R_1CH_2\overset{O}{\overset{\diagup\diagdown}{CH-CH_2}} + R_2OH \rightarrow R_1CH_2-\overset{OH}{\underset{|}{CH}}-CH_2OR_2 \qquad (4)$$

homopolymerization, hydroxyl groups are present in the impure TGDDM (MY720) as α-glycol groups and further hydroxyl groups can form in the 177–300 °C range as a result of isomerization and/or oxidation of the epoxide groups. The presence of the tertiary nitrogen NR_3 groups in the TGDDM molecule should enhance the epoxide-hydroxyl (E-OH) reaction. The hydroxyl groups present in MY720 accelerate the rate of epoxide consumption compared to pure TGDDM, which indicates that the E-OH reactions rather than epoxide-epoxide (E-E) reactions are the predominant chain extension reactions for TGDDM.

A number of workers have reported epoxides isomerize to allylic alcohols (5) [30, 35–38, 40, 41, 51–56]. Epoxide isomerization to

$$R-CH_2-\overset{O}{\overset{\diagup\diagdown}{CH-CH_2}} \rightarrow R-CH_2-\overset{OH}{\underset{|}{C}}=CH_2 \quad \text{or} \quad R-CH=CH-CH_2OH$$
$$(5)$$

allylic alcohols and also to an aldehyde (6) is consistent with the simultaneous appearance of hydroxyl and carbonyl groups upon epoxide consumption. The absence of a methyl

$$R-CH_2-\underset{\underset{O}{\diagdown\!\diagup}}{CH-CH_2} \rightarrow R-CH_2CH_2-CHO \qquad (6)$$

group in the IR spectra indicates isomerization to an aldehyde rather than a ketone is preferred.

In addition to isomerization some of the epoxide groups could be oxidized to hydroxyl aldehyde and then carboxylic acid groups (7)[57]. These oxidation reactions would also be consistent with our FTIR observations.

$$R-CH_2-\underset{\underset{O}{\diagdown\!\diagup}}{CH-CH_2} \xrightarrow{O_2} R-CH_2\underset{OH}{\overset{|}{C}}H-C\underset{H}{\overset{\diagup\!O}{\diagdown}} \xrightarrow{O_2} R-CH_2\underset{OH}{\overset{|}{C}}H-C\underset{OH}{\overset{\diagup\!O}{\diagdown}} \qquad (7)$$

Hence, the most plausible explanation of our FTIR observations of the simultaneous appearance of hydroxyl, carbonyl and ether groups upon TGDDM epoxide consumption is epoxide isomerization and/or oxidation followed by epoxide-hydroxyl chain extension reactions.

At the higher temperature range of our studies, 225–300 °C, degradation reactions of the polymerized network occur. Previous studies on the degradation of epoxies indicate dehydration will be the principal degradation mechanism in the 225–300 °C temperature range over time periods of hours [36, 58-63] which is consistent with the observed decrease in hydroxyl band intensity with increasing temperature in this range.

5.3 TGDDM-DDS Cure Reactions

In this Sect. we report FTIR studies of the cure reactions of TGDDM-DDS-$BF_3:NH_2C_2H_5$ epoxy systems as a function of cure conditions (100–300 °C) and DDS (0–35 wt.-%) and $BF_3:NH_2C_2H_5$ (0–5 wt.-%) concentrations. We monitored by FTIR difference spectra the disappearance of the epoxide and P.A. groups and the appearance of the S.A., ether and hydroxyl groups. The I.R. bands at 1630 and 3410 cm^{-1} were assigned to the P.A. and S.A. groups respectively and their intensities were normalized relative to the 1516 cm^{-1} band associated with the phenyl group. The intensity of S.A. group was further normalized against the O=S=O band from the DDS structure ($I_{NH:SO_2}$) which we assume does not change in intensity during cure. This normalization allows comparison of S.A. changes independent of the effects of DDS concentration. We, also, normalized the OH and ether IR band intensities relative to the intial epoxide band intensity by the following expression, illustrated in this case for the OH band intensity (R_{OH}) where IOH_T, $IOH_{23°C}$, $IE_{23°C}$

$$R_{OH} = \frac{IOH_T - IOH_{23°C}}{IE_{23°C} \times (M.F.)_{TGDDM}} \qquad (8)$$

and (M.F.) TGDDM are the normalized OH band intensities at cure temperature T and 23 °C, epoxide band intensity at 23 °C and the mole fraction of epoxide groups relative to pure TGDDM, respectively. The disappearance of the epoxides and P.A.'s in % and the appearance of the S.A.'s, OH's and ethers in the form of $I_{NH:SO_2}$, R_{OH} and R_{ether} values respectively are plotted in Fig. 14 as a function of cure time at 177 °C for a TGDDM-DDS (25 wt.-% DDS) (0.4 wt.-% $BF_3:NH_2C_2H_5$) epoxy system. After 210 min at 177 °C, 85% of the epoxide groups are consumed. At the latter stages of cure the epoxide consumption rate is evidently hindered by steric restrictions. The P.A.-E reaction dominates the early stages of cure. After 30 min at 177 °C, 95% of P.A. groups and 45% of the epoxide groups have reacted; 28% of these epoxide groups are consumed by the P.A.-E reaction. From FTIR and DSC studies of the cure of a non-$BF_3:NH_2C_2H_5$ catalyzed TGDDM-DDS epoxy at 177° Moacanin et al. [64] and Gupta et al. [22] have concluded the P.A.-E reaction dominates the early stages of cure. Also, chemical titration and liquid chromatography studies [65] indicate epoxide consumption is linear with time during the early stages of cure which is also consistent with the predominance of the P.A.-E reaction.

In the 30–90 min cure time range there is a ~30% increase in the R_{OH} intensity, despite 95% of the P.A.'s being consumed after 30 min. This increase could be caused

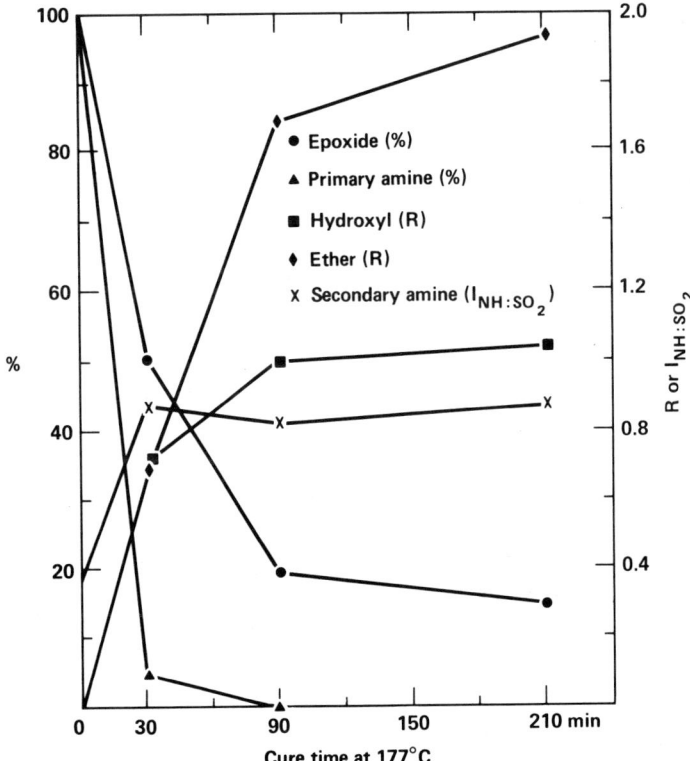

Fig. 14. Epoxides (●) and P.A.'s (▲) (% of unreacted groups), R_{OH} (■) R_{ether} (♦) $I_{NH:SO_2}$ (X) vs. cure time at 177 °C for TGDDM-DDS (25 wt.-% DDS) (0.4 wt.-% $BF_3:NH_2C_2H_5$) epoxy

by ~25% of the S.A.'s reacting with 7% of the epoxides via S.A.-E addition reactions. The $I_{NH:SO_2}$ intensity decrease associated with the S.A.-E reaction would be difficult to detect because the P.A.-E reaction simultaneously causes an increase in this intensity. After 90 mins of cure the R_{OH} and $I_{NH:SO_2}$ values remain constant with increasing cure time which indicates the S.A.-E reaction does not occur during the later stages of cure probably because of network steric restrictions. The S.A.-E reaction rate constant for aromatic amines with structures similar to DDS, such as methylene dianiline, have been reported to be 7–12 times slower than for the P.A.-E reaction [66].

The increase in the R_{ether} throughout the cure is associated with the E-OH reaction which consumes ~50% of the total epoxides.

The cure reactions of TGDDM-DDS (25 wt.-% DDS) epoxy were monitored as a function of $BF_3:NH_2C_2H_5$ concentration (0–5 wt-%) in the 100 177 °C cure temperature range. The cure reactions were accelerated with increasing $BF_3:NH_2C_2H_5$ concentration and shifted to lower cure temperatures.

The cure and degradation reactions of TGDDM-DDS epoxies were also monitored by FTIR in the 177–300 °C as a function of DDS concentration (0–35 wt.-%). Three series of epoxies were studied containing (i) 0 wt.-% $BF_3:NH_2C_2H_5$, (ii) 0.4 wt.-% $BF_3:NH_2C_2H_5$ and (iii) 0.4 wt.-% $BF_3:NH_2C_2H_5$-acetone mixed.

In Fig. 15 we illustrate data for a TGDDM-DDS epoxy series (0 wt.-% $BF_3:NH_2C_2H_5$) in which the disappearance of the epoxides and P.A.'s in % and the appearance of the S.A.'s, OH's and ethers in the form of $I_{NH:SO_2}$, R_{OH} and R_{ether} values, respectively, are plotted as a function of DDS concentration and cure conditions. Similar plots were generated for the other two $BF_3:NH_2C_2H_5$ epoxy systems.

All P.A. groups are consumed at 177 °C after 2.5 h cure for all TGDDM-DDS epoxy systems studied that contain < 35 wt.-% DDS. For TGDDM-DDS (35 wt.-% DDS) epoxies, however, ~5% of P.A.'s remain unreacted even at 300 °C, which suggests that these unreacted groups become inaccessible to epoxide groups during the later stages of cure because of network topography constraints.

In Fig. 16, plots of epoxide consumptions at 177 °C after 2.5 h vs. wt.-% DDS for all TGDDM-DDS systems studied are illustrated. From these data and similar plots at higher cure temperatures we conclude: (i) At 177 °C for a 2.5 h cure for TGDDM-DDS (0.5 wt.-% DDS) epoxies, the $BF_3:NH_2C_2H_5$ catalyst enhances epoxide consumption 2–4 times. (ii) At cure temperatures in the 200–225 °C range we found no definitive evidence that the $BF_3:NH_2C_2H_5$ catalyst enhances epoxide consumption for all systems studied, which suggests the principal catalytic species, the $BF_4^-NH_3^+C_2H_5$ salt, is deactivated in this temperature regime. (iii) At 177 °C for TGDDM-DDS (15–35 wt.-% DDS) epoxies the $BF_3:NH_2C_2H_5$ catalyst enhances epoxide consumption, but does not always produce a fully cured system. The large scatter in these epoxide consumptions, 60–100%, from a number of experiments for the TGDDM-DDS (15–35 wt.-% DDS) epoxies, suggests there is a considerable variability in the $BF_3:NH_2C_2H_5$ catalytic activity. From our DSC and NMR studies of $BF_3:NH_2C_2H_5$ catalytic activity, we attribute the epoxide consumption data scatter to variability in inherent $BF_3:NH_2C_2H_5$ chemical composition and its particle size distribution within the epoxy, and also non-uniform mixing of this small concentration of catalyst in the epoxy. (iv) From two completely independent series of acetone and non-acetone mixed epoxies we found no difference in epoxide consump-

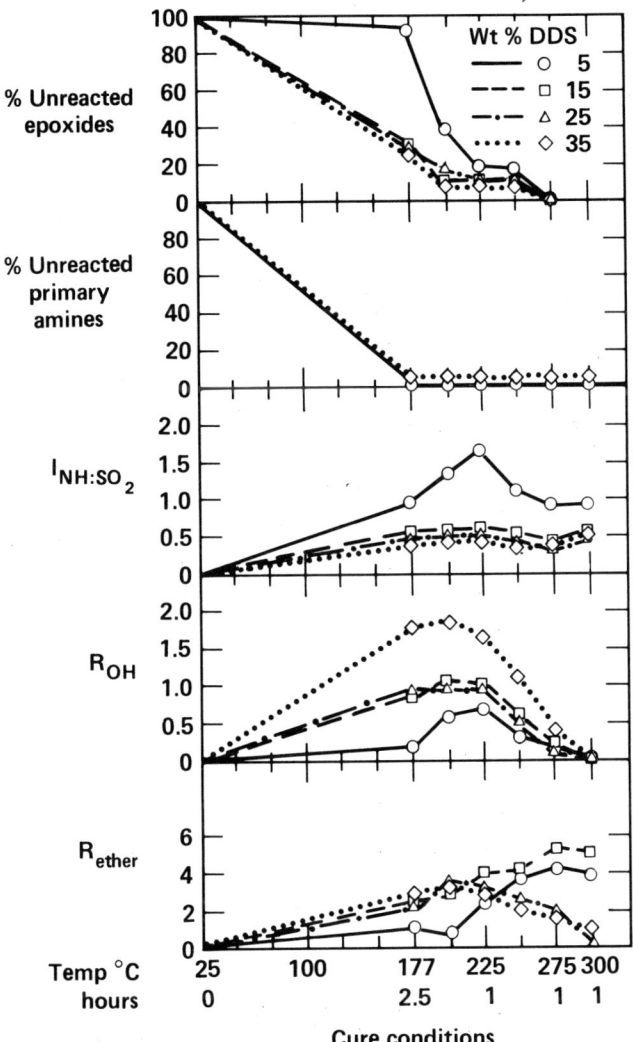

Fig. 15. % of unreacted epoxide and P.A. groups, $I_{NH:SO_2}$, R_{ether} and R_{OH} values vs. cure conditions in the 177–300 °C range and DDS concentration for TGDDM-DDS (0 wt.-% $BF_3:NH_2C_2H_5$) epoxies

tion within experimental scatter. (v) The epoxide consumption is enhanced with increasing DDS concentration to a greater extent than that associated with 100% completed P.A.-E and 50% completed S.A.-E reactions (Fig. 16). The hydroxyl products of the P.A.-E and S.A.-E reactions enhance epoxide consumption via E-OH reactions. From their DSC studies Mijovic er al.[67] suggest the cure reactions of TGDDM-DDS epoxies are autocatalytic.

The $I_{NH:SO_2}$ intensity after 2.5 hrs cure at 177 °C decreases with increasing DDS concentration for the three TGDDM-DDS systems studied, as illustrated in Fig. 17. For a TGDDM-DDS (25 wt.-% DDS) (0 wt.-% $BF_3:NH_2C_2H_5$) epoxy the $I_{NH:SO_2}$

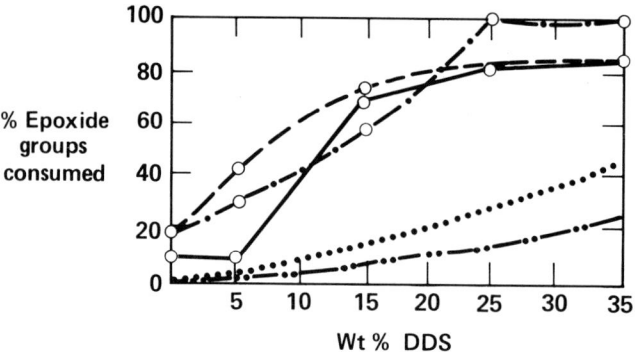

Fig. 16. % of epoxide groups consumed at 177 °C after 2.5 hrs vs. wt.-% DDS for (i) TGDDM-DDS (0 wt.-% $BF_3:NH_2C_2H_5$) (———); (ii) TGDDM-DDS (0.4 wt.-% $BF_3:NH_2C_2H_5$) non-acetone mixed (----------); (iii) TGDDM-DDS (0.4 wt.-% $BF_3:NH_2C_2H_5$) acetone mixed (—·—·—·); (iv) 100% completed P.A.-E reaction (..........) and (v) 50% completed S.A.-E reaction (—··—··—)

intensity is 30–40% lower than the maximum intensity observed at lower DDS concentrations, thus suggesting at least ~1/3 of the S.A. groups do react with epoxide groups in commercial prepregs under standard cure conditions.

The $I_{NH:SO_2}$ intensity increases in the 177–225 °C region, despite the completion of the P.A.-E cure reaction that would result in additional S.A. groups. This $I_{NH:SO_2}$ intensity increase is more pronounced for lower DDS concentration systems. From vaporization gas chromatography/mass spectroscopy studies Grayson and Wolf[68] report propenal is the principal degradation product of a TGDDM-DDS (21 wt.-% DDS) (0 wt.-% $BF_3:NH_2C_2H_5$) epoxy system in the 125–215 °C region. Propenal was not observed, however, when MY720 was heated in the same temperature region[69],

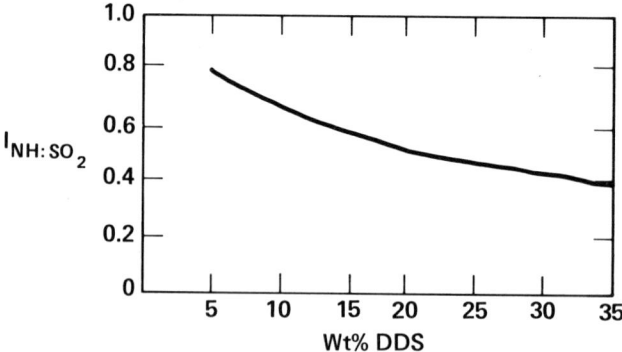

Fig. 17. $I_{NH:SO_2}$ intensity after 2.5 hrs at 177 °C vs. DDS concentration for TGDDM-DDS epoxies

which discounts propenal was formed by unimolecular decomposition of unreacted epoxide groups (9).

$$R-N\begin{matrix}CH_2-CH-CH_2\\ \diagdown O\diagup \\ CH_2-CH-CH_2\\ \diagdown O\diagup\end{matrix} \rightarrow R-N\begin{matrix}H\\ \\ CH_2-CH-CH_2\\ \diagdown O\diagup\end{matrix} + CH_2=CH-C\begin{matrix}H\\ \diagdown\\ O\end{matrix} \quad (9)$$

Propenal and the S.A. group formation must form from decomposition of the P.A.-E, S.A.-E, and E-OH cure reaction products. However, because we observe the S.A. group formation is most prevalent for low DDS concentration TGDDM-DDS epoxies, which contain a higher concentration of E-OH cure reaction products, then we suggest propenal and S.A. group formation most likely occurs by decomposition of non-cyclized (10) and cyclized (11) E-OH reaction products. (see next Sect. Rates and Chemistry of Cure Reactions)

$$-N\begin{matrix}R_1\\ \\ CH_2-CH(OH)-CH_2O-R_2\end{matrix} \rightarrow -N\begin{matrix}R_1\\ \\ H\end{matrix} + CH_2=CH-C\begin{matrix}H\\ \diagdown\\ O\end{matrix} \quad (10)$$

$$+ R_2H$$

$$-N\begin{matrix}CH_2-CH--\\ \diagdown O\\ CH_2-CH\diagdown\\ CH_2OH\end{matrix} \rightarrow -N\begin{matrix}CH_2-CH_2(OH)--\\ \\ H\end{matrix} + CH_2=CH-C\begin{matrix}H\\ \diagdown\\ O\end{matrix} \quad (11)$$

Similar decomposition mechanisms have been proposed by David [70] and Grayson and Wolf [68].

The hydroxyl band intensities, plotted in the form of R_{OH} values decrease with increasing temperature above 200 °C for epoxies containing 15–35 wt.-% DDS, which indicates these epoxies dehydrate in this temperature regime. For epoxies containing 0–10 wt.-% DDS, dehydration is not evident until above 250 °C because isomerization of unreacted epoxide groups causes an increase in the concentration of OH groups in the 177–250 °C temperature range. The R_{OH} value for a 177 °C, 2.5 h cure increases with increasing DDS concentration, as illustrated in Fig. 18, because of the enhancement of the P.A.-E and S.A.-E cure reactions with increasing DDS concentration. No significant differences in the dehydration processes were observed for the three TGDDM-DDS epoxy series that were studied.

Ether formation can occur from the E-OH reaction. In Fig. 19a, the R_{ether} value for a 177 °C, 2.5 h cure increases with increasing DDS concentration in the 0–15 wt.-% DDS range because of the enhanced availability of OH groups from the P.A.-E and S.A.-E addition reactions. For >15 wt.-% DDS concentrations, ether formation is not enhanced with increasing DDS concentrations because the competing epoxide-amine addition reactions consume the available epoxide groups. At 0 wt.-% DDS, the $BF_3:NH_2C_2H_5$ catalyst enhances ether formation ~5 times, but in >0 wt.-% DDS epoxies, the $BF_3:NH_2C_2H_5$ catalyst has no detectable effect on ether formation for a 177 °C, 2.5 h cure. At higher cure temperatures from 177 to 300 °C R_{ether} values

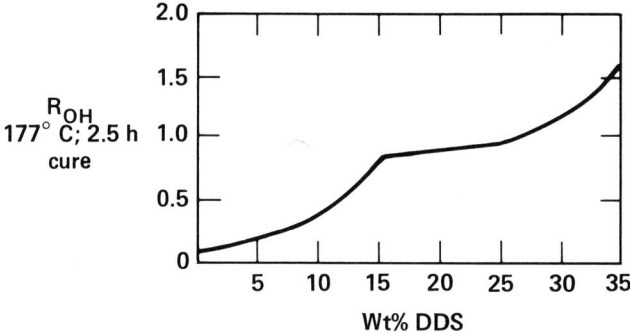

Fig. 18. R_{OH} values after 177 °C–2.5 hr cure vs. DDS concentration for TGDDM-DDS epoxies

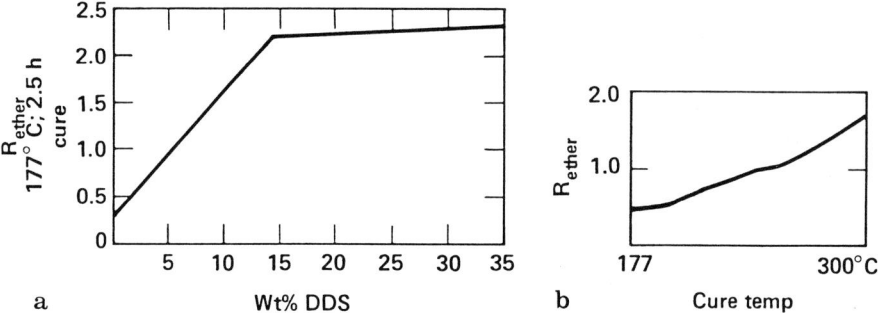

Fig. 19a and b. (a) R_{ether} values after 177 °C–2.5 hr cure vs. DDS concentration and (b) R_{ether} values vs. cure temperature for TGDDM-DDS epoxies

increase with increasing cure temperature, Fig. 19b, as a result of dehydration, and/or network oxidation and resultant formation of ether crosslinks.

5.4 Rates and Chemistry of Cure Reactions

The reaction rates of the three principal cure reactions of TGDDM-DDS epoxies are as follows:

$$d(PA\text{-}E)/dt = k[PA][E]$$
$$d(SA\text{-}E)/dt = 0.1k[SA][E]$$
$$d(E\text{-}OH)/dt = 0.1k[OH][E]$$

The rate constants for the S.A.-E and E-OH reactions are 10 times slower than the rate constant for the P.A.-E reaction. From FTIR and DSC studies Moacanin et al. [64] report the E-OH reaction is 10 times slower than the P.A.-E reaction at 177 °C. We previously mentioned that the S.A.-E rate constant for aromatic amines similar to DDS, such as methylene dianiline have been reported to be 7–12 times slower than the P.A.-E reaction [66].

In Fig. 20a plot of the % of epoxides consumed by each cure reaction vs. total percent epoxide consumed for a TGDDM-DDS (25 wt.-% DDS) epoxy cured at 177 °C is illustrated. The P.A.-E reaction dominates the early stages of cure, until all the P.A. is depleted. The E-OH reaction dominates the later stages of cure.

The P.A.-E reaction is illustrated in Fig. 21, with the sites for further cure reactions via the (i) E-OH and (ii) S.A.-E reactions indicated by arrows. The E-OH and S.A.-E reactions can occur (a) intermolecularly to form crosslinks or (b) intramolecularly

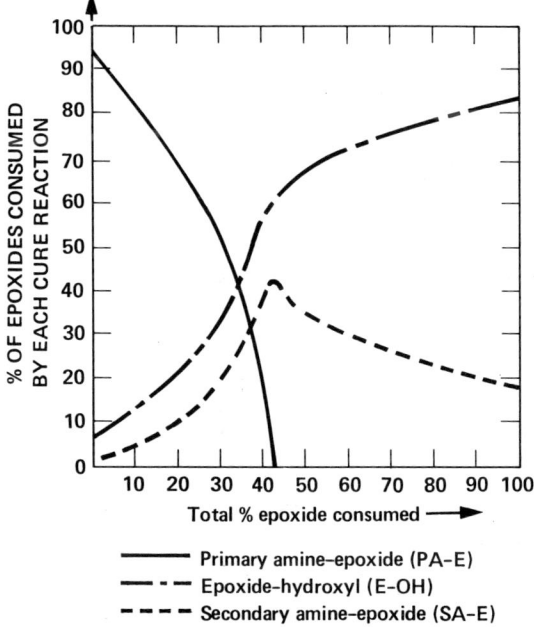

Fig. 20. % epoxide consumed by each cure reaction vs. total % epoxide consumed

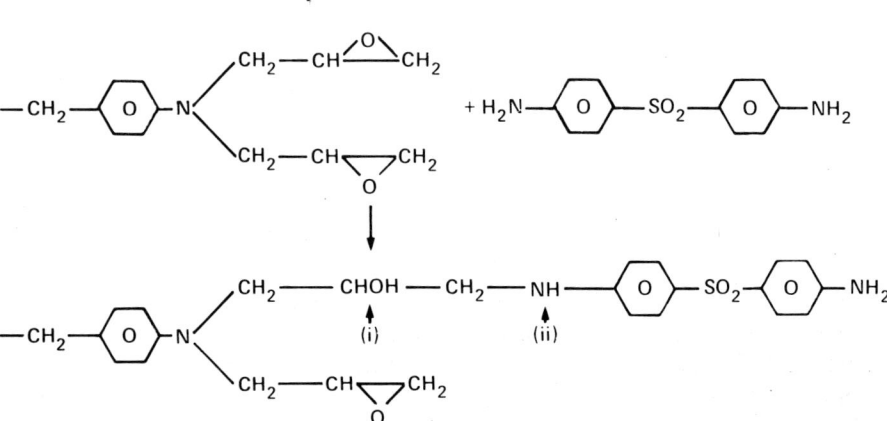

Fig. 21. Primary amine-epoxide reaction ((i) and (ii) — sites for E-OH and SA-E reactions respectively

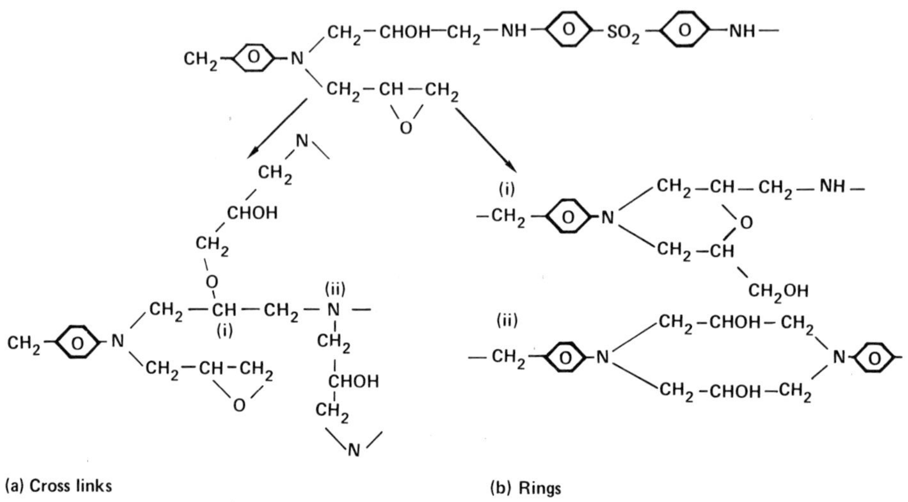

Fig. 22a and b. (i) Epoxide-hydroxyl and (ii) secondary amine-epoxide reactions that form (**a**) intermolecular crosslinks and (**b**) intramolecular rings

to form noncrosslinked internal rings, as illustrated in Fig. 22. Molecular modeling studies for low viscosity (η) cure conditions indicate $\sim 75\%$ of the E-OH and S.A.-E reaction occur by ring formation. (Higher η cure conditions would favor a higher percentage of ring formation.)

5.5 Prepreg Processing Viscosity

To produce reproducible composites from prepregs, it is important for the prepreg to exhibit reproducible η-temperature-time profiles. Generally, prepregs are isothermally cured for a period of time at a lower temperature (i.e., 130 °C) prior to final cure at 177 °C. A minimum in η will occur during this isothermal cure, and pressure is applied to the composite at this time to remove excess resin and to produce a void-free composite. The η of a liquid normally decreases with increasing temperature, but as the cure reactions occur the η will increase with increasing temperature and time. The magnitude and shape of the η-temperature-time profile of the prepreg will depend on (1) the initial η of MY720, (2) the temperature-time mixing conditions of the epoxy components and their introduction onto the C-fibers and (3) the melting point, catalytic activity and the distribution of the BF_3:amine catalyst within the prepreg. The size and distribution of the BF_3:amine catalyst particles within the prepreg is probably difficult to control and highly variable. Also, any recrystallization of the DDS in the prepreg changes the concentration of P.A. available for P.A.-E reactions below the DDS melting point at 162 °C. The rate of the P.A.-E cure reaction plays the predominant role in the rate at which the η increases with temperature and time.

6 Physical Structure

6.1 Introduction

The principal physical structural parameters that control the modes of deformation and failure and mechanical response of epoxies are (1) macroscopic inhomogenieties such as microvoids or concentrations of unreacted monomer, (2) the glassy-state free volume and (3) the crosslinked network structure characteristics.

6.2 Macroscopic Inhomogeneities

Microvoids can result when air, moisture or other low molecular weight material is trapped in the system during cure and subsequently vaporized and possibly eliminated from the epoxy during postcure. The low molecular weight material results from either inhomogeneous mixing of the starting components or from aggregation, i.e. crystallization, of unreacted constituents. In polyamine-cured DGEBA epoxies, crystals of DGEBA epoxide monomer trapped in the partially cured resin at 23 °C can produce microvoids by melting and volatilizing under certain post cure conditions [71]. Thermal-anneal, moisture-sorption and mechanical property studies also indicate that in TGDDM-DDS epoxies, the melting and volatilization of unreacted DDS crystallites during cure produce microvoids [72].

6.3 Free Volume

The mechanical properties of epoxies exhibit a free-volume dependence as a function of thermal history [47, 73, 74]. Changes in free volume, or local order, in the glassy state can occur as a result of the extension to temperatures below T_g of packing changes associated with the liquid state. The liquid-volume temperature plot extrapolated to below T_g in Fig. 23 represents the lower free-volume, equilibrium state of the glass. The time necessary to achieve the equilibrium state at a given temperature below T_g depends on the glassy-state mobility. Below a specific temperature, the glassy-state mobility is too small to allow any changes in free volume. A decrease in free volume that occurs in the glassy state results in inhibition of the flow processes that occur during deformation and a more brittle mechanical response. Rapid cooling from above T_g, however, produces a glass with a larger free volume.

In addition to thermal history the free volume of epoxies also depends on the packing ability of the epoxide and amine structure and the geometric constraints imposed on the segmental packing by the crosslinked network geometry. Molecular model studies of DGEBA-T403 epoxies indicate that as the crosslink density of the network increases, the packing efficiency decreases. From compressibility studies, Findley and Reed [75] report lower crosslinked epoxies are more compact. Also, diffusion studies of H_2 and O_2 in crosslinked glasses indicate that the permeabilities will exhibit maxima at the highest crosslink density if the units that form the network inhibit close packing because of steric restrictions [76-78]. In our experimental studies on DGEBA-T403

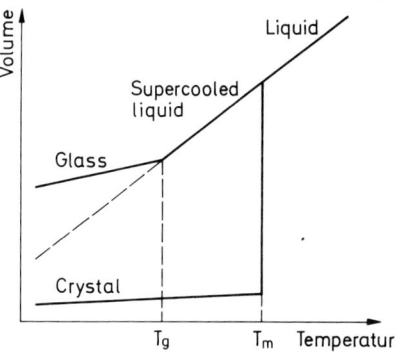

Fig. 23. Volume-temperature plot for a polymer

epoxies we observed that these glasses exhibited a minimum in density and modulus and, therefore, maximum in free volume at the highest crosslink density. Furthermore, from dynamic mechanical measurements we observed the highest crosslink density DGEBA-T403 epoxy exhibited the largest secondary glass transition intensity. This observation implies the network segments responsible for the glassy state molecular motion associated with the secondary glass transition possess the greatest mobility and, therefore, free volume at the highest crosslink density.

6.4 Network Structure

The critical network structural parameters that control the mechanical performance of epoxies are macroscopic heterogeneities in crosslink density and the network topography on the molecular level.

Epoxies can form networks with heterogeneous crosslink density distributions in the 6 to 10,000 nm size range [73, 79–122]. The formation of a heterogeneous rather than a homogeneous system depends on polymerization conditions, i.e. temperature, solvent, chemical composition and mixing of the starting materials. The high crosslink density regions have been described as agglomerates of colloidal particles [86, 87] or floccules [91] in a lower molecular weight interstitial fluid. Funke [123, 124] suggested a number of factors that can be responsible for heterogeneous network formation: 1) difference between the reactivities of different functional groups, 2) unreacted functional groups, 3) intramolecular cyclization reactions and 4) phase separation. Phase separation in the form of microgelation is generally believed to be the primary mechanism for the formation of heterogeneous crosslinked networks. Solomon et al.[90] originally suggested that a two-phase system is produced by microgelation prior to the formation of a macrogel. Kenyon and Nielsen [94] indicated that the highly crosslinked microgel regions are loosely connected during the latter stages of the curing process. Karyakina et al. [104] suggested that microgel regions originate in the initial stages of polymerization from the formation of microregions of aggregates of primary polymer chains. Luettgert and Bonart [112] discussed the morphology of epoxies in terms of the relative rates of microgel formation and the subsequent growth rate of these gel particles. At low cure temperatures, only a small number of gel particles are nucleated and, hence, large nonhomogeneities are produced; at higher cure temperatures, the rate of nucleation of microgel particles is faster, and larger numbers are produced, which, therefore, limits their growth in size. More recently, Bell [121] has postulated and shown that local regions of higher than average crosslink density can be attributed to inadequate mixing of the epoxy starting components. There have been no studies of the mutual solubilities of epoxy starting components as a function of temperature and composition.

The majority of the evidence for heterogeneous regions of crosslink density in epoxies is derived from electron microscope investigations. These microscopy studies involve carbon-platinum replication of etched and nonetched free surfaces and fracture surfaces. However, artifacts can often result from replication techniques [26]. More confidence can be placed in bright-field transmission electron microscopy studies of the morphology of thin epoxy films strained directly in the electron microscope [110]. The morphology of such thin films, however, may not be representative of the bulk. However, if the heterogeneous regions are sufficiently large as in the case of commercial TGDDM-DDS (25 wt.-% DDS, 0.4 wt.-% $BF_3:NH_2C_2H_5$) epoxies such regions can be detected directly by optical microscopy [27, 111].

Small angle x-ray scattering (SAXS) studies have the potential to monitor crosslinked network morphologies. Recent studies [125] conducted for us at the National Scattering Center, Oak Ridge did not detect any heterogeneous regions that could be associated with crosslink density variations in a number of epoxies utilized as high performance composite matices. However, if such regions were detected by SAXS it would be necessary to understand the relationship between the density and degree of crosslinking of the network to interpret such observations. This requires an understanding of the network structure on the molecular level.

The network parameters that can affect the mechanical response of a crosslinked epoxy are the network defects and topography.

Network defects in the form of unreacted groups serve as sites for crack initiation and propagation. When such defects are non-randomly distributed within the network a nodular morphology will be observed upon fracture or chemical etching of the bulk network.

The ability of a crosslinked network to deform depends on the glassy-state free volume and network extensibility. The extensibility of the network segments rather than the molecular weight between crosslinks, M_c determines the ultimate network properties. From the theory of rubber elasticity, the maximum network extensibility is directly related to M_c by the expression, $\lambda_b \alpha M_c^{1/2}$ [126]. In highly crosslinked systems, however, the extensibility of the network segments between crosslinks depends on specific rotational isomeric configurations of the segments and the perturbation of these configurations by geometric network constraints. The network topography of the highly crosslinked networks will also significantly affect the network extensibility. To describe the network topography of epoxies, we first of all investigate the basic ring structures of the network by molecular models. These studies indicate which ring structures are sterically possible. The ring structures for DGEBA-T403 epoxies are illustrated in Fig. 24. Molecular and computer modeling studies indicate the deformability of these basic ring structures depends, 1) the extensibility of their sides; 2) the flexibility of their internal angles; and 3) their ability to undergo co-operative deformation with their interconnected neighbors. Their co-operative deformation is controlled by the regularity of the network topography which is determined by the

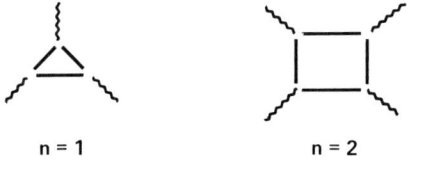

Series (A) Sides of rings consist of (1 DGEBA segment + 2 arms of the T403 molecule)n

Where n = 1, 2, 3, . . .

n = 1   n = 2

Series (B) Sides of rings consist of (2 DGEBA segments + 2 arms of T403 molecule)n

Where n = 1, 2, 3, . . .

n = 1   n = 2

Series (C) Sides of rings consist of (2 DGEBA segments + n DGEBA segments)

Where n = 1, 2, 3, . . .

n = 1   n = 2

T403

DGEBA ———

Fig. 24. Ring structures in DGEBA-T403 epoxy networks

different geometries of the basic rings and their orientation relative to one another. The more regular networks consist of interconnected rings of similar size and shape. However, the more irregular networks consist of rings with a variety of geometries that will develop overstrained segments at lower extensibilities. The fully crosslinked DGEBA-T403 epoxy network is relatively irregular as shown in Fig. 25. The deformability of the basic ring structures will also depend on the direction of the applied stress field relative to the ring structure. This orientation factor will significantly affect the networks that consist of regions of regularly oriented and interconnected rings.

The distribution in network segmental extensibilities controls failure initiation and propagation of the network. The segments with the lowest extensibilities at any given

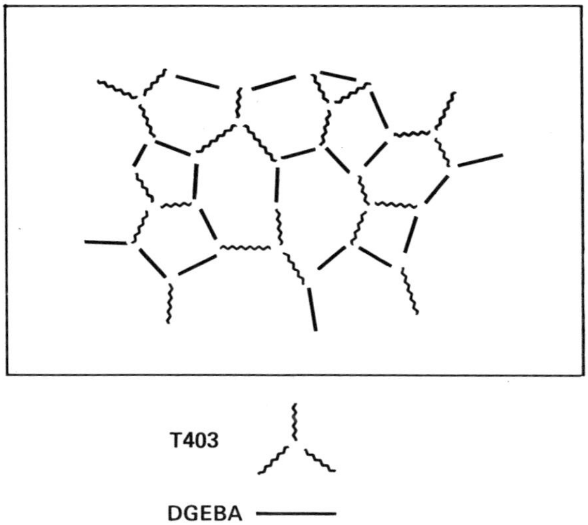

Fig. 25. Network structure of fully reacted stoichiometric DGEBA-T403 epoxy

time in the failure process will carry a significant portion of the load and will undergo chain scission resulting in a process of progressive scission of the least extensible segments. Discrete rather than random distributions in the network extensibilities of epoxies exist because of the chemical nature of epoxy network formation.

7 Deformation and Failure Modes

Epoxies can undergo plastic flow prior-to or during failure if there is sufficient free volume and time for network segmental reorientation to occur. A variety of evidence shows that plastic flow can occur under tensile loads in crosslinked amine-cured epoxy glasses [1, 127]. This flow can occur by homogeneous deformation, or inhomogeneously via crazing and/or shear banding. Cracks are often initiated in these glasses by a crazing process.

The tensile fracture topographies of un-notched amine-cured epoxies exhibit three characteristic regions: 1) a course initiation region; 2) a slow crack growth, smooth region; and 3) a fast crack growth, rough region [1, 71–73, 110–127]. The coarse initiation region, that can contain microvoids and/or fractured fibrils, is a result of crack propagation through coarse crazes and/or shear bands. The crack then imposes a higher stress field on the craze or flaw tip which produces a small plastic zone that results in a smooth fracture topography. Whether flow in this small plastic zone at the crack tip occurs by shear yielding or yielding under normal stresses is difficult to ascertain experimentally and will depend on the stress fields imposed on the epoxy immediately ahead of the crack tip. The area of the mirror fracture topography is a measure of the ability of the polymeric glass to undergo plastic flow at the propagating crack tip.

In Fig. 26, we schematically illustrate four stages of failure in epoxies under an increasing tensile load. In each stage we document the craze/crack structure, the stress at the craze/crack surface and the resultant fracture topography.

In addition to the fracture topography observations in the slower crack propagation regions of epoxies a number of additional observations also indicate permanent molecular flow can occur in amine-cured epoxy networks in the glassy state.

Amine-cured epoxies can exhibit macroscopic tensile yield stresses in the glassy state [47, 73]. Such yield stresses exhibit similar free volume dependencies as a function of thermal history as non-crosslinked glasses such as polycarbonate [47, 73, 128]. From Eyring's theory of stress-activated viscous flow in polymers [129] the strain rate and temperature dependencies of the epoxy yield stresses produce activation volumes associated with chain segment jumps similar in magnitude to those volumes associated with flow in non-crosslinked glasses [47]. Furthermore, many of the amine-cured epoxies we have studied exhibit ultimate elongations in the 10–30% range in the glassy state [1, 47, 72, 73].

Transmission electron microscopy (TEM) and birefringence studies of strained and/or fractured epoxies have revealed more direct experimental evidence that molecular flow can occur in these glasses. Films of DGEBA-DETA (~ 11 wt.% DETA) epoxies, ~ 1 μm thick, were strained directly in the electron microscope and the deformation processes were observed in bright-field TEM [73, 110]. Coarse craze fibrils yielded inhomogeneously by a process that involved the movement of indeformable 6–9 nm diameter, highly crosslinked molecular domains past one another. The material between such domains yielded and became thinner as plastic flow occurred.

We have monitored the deformation processes of a number of amine-cured DGEBA epoxies under polarized light [1, 127]. The most thorough study was conducted on DGEBA-T403 epoxies at 23 °C as a function of strain level and epoxide:amine ratio. The birefringent patterns of a DGEBA-T403 stoichiometric epoxy (47 phr T403)

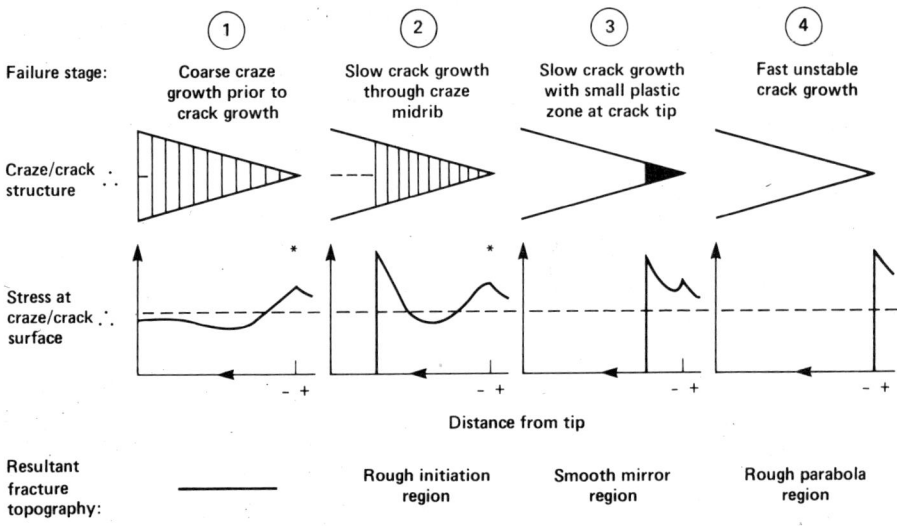

Fig. 26. Microscopic failure processes in epoxies in tension

as a function of strain are shown in Fig. 27. At strains <2.5% only elastic deformations occur and homogeneous color changes produced under polarized light in this strain region disappear instantaneously upon removal of the load. In the 2.5–4% strain region, homogeneous elastic and plastic deformations occur, and upon removal of the load the homogeneous plastic deformation does not relax out, which results in a permanent homogeneous color change in the unstrained epoxy when viewed under polarized light. Above 4% strain, the plastic deformation becomes increasingly inhomogeneous with increasing strain as indicated by the development of birefringent fringes. Ultimately, a local region of high strain develops in the sample in the form of a diffuse shear band and a neck develops in this region. The orientation of the birefringent fringes are initially affected by the network defect topography but ultimately they orient as a shear band at $\approx 45°$ to the direction of the applied load. Fracture occurs in the high strain, shear band region. Occasionally, more than one diffuse shear band will develop in the sample.

For the off-stoichiometric DGEBA-T403 epoxies that are in the 30–75 phr T403 range, birefringent-strain studies indicate the following trends in the deformation processes for epoxies that increasingly are removed further from the stoichiometric T403 concentration: 1) plastic flow is initiated and also becomes inhomogeneous at lower strains; 2) the regions of high strain are less oriented and their associated shear bands are less well developed upon crack propagation and ultimate failure through such regions. DGEBA-T403 epoxies in the 20–30 phr T403 range that exhibit ultimate elongations in the 20–70% range deform only in a homogeneous plastic fashion.

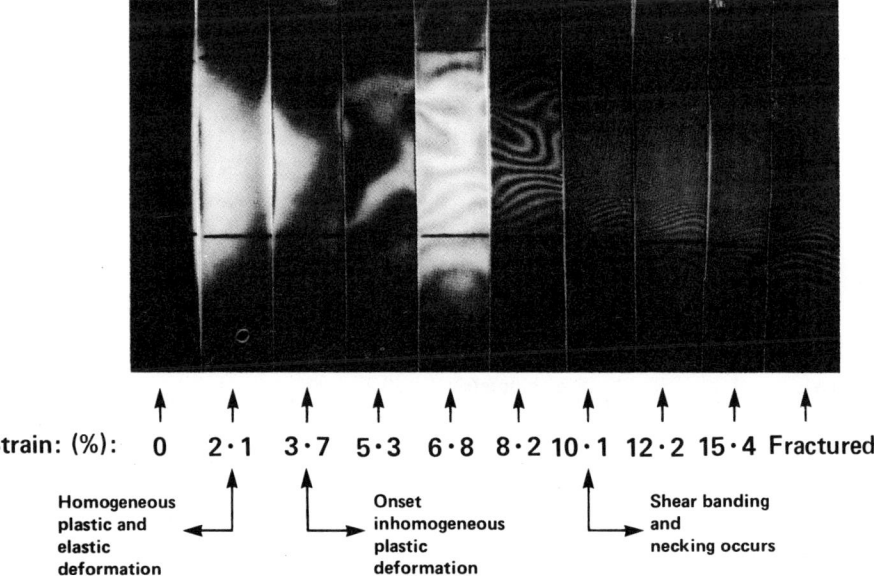

Fig. 27. Birefringent deformation processes in DGEBA-T403 (47 phr T403) epoxy as a function of strain at 23 °C, under polarized light

8 Structural Parameters that Control Mechanical Properties

The flexibility and extensibility of a crosslinked epoxy network are determined by the available glassy-state free volume. If the free volume is insufficient to allow network segmental extensibility via rotational isomeric changes then the brittle mechanical response of the epoxy glass is not controlled by the network structure but rather by macroscopic defects such as microvoids. For epoxies with sufficient free volume that allows plastic network deformation the mechanical response is controlled by the network structure.

Studies on DGEBA-T403 epoxies[1] reveal the magnitudes of the yield stress, tensile strength (which is controlled by the flow properties of the epoxy) and Young modulus are determined by the glassy-state packing and free volume. These mechanical properties follow the same trends as the density as a function of T403 concentration. The density of the DGEBA-T403 epoxies decreases with increasing T403 content with a slight minimum superimposed on such a downtrend at stoichiometry because of the constraints of the crosslinks on the network packing efficiency.

The ultimate extensibility of an epoxy glass depends on the network structure extensibility and its associated defects. We found the ultimate glassy-state extension for DGEBA-T403 epoxies is a maximum at stoichiometry and decreases either side of stoichiometry as more defects are indroduced into the network in the form of unreacted groups (Fig. 28). The presence of unreacted groups and their associated unconnected, non-load bearing segments results in higher loads being imposed under stress on those segments that are incorporated into the network. The higher segmental stresses enhance stress-induced chain scissions at lower network extensibilities. The critical concentration of chain scissions associated with crack initiation will be attained at increasingly lower network extensibilities as the initial concentration of unreacted groups in the network increases.

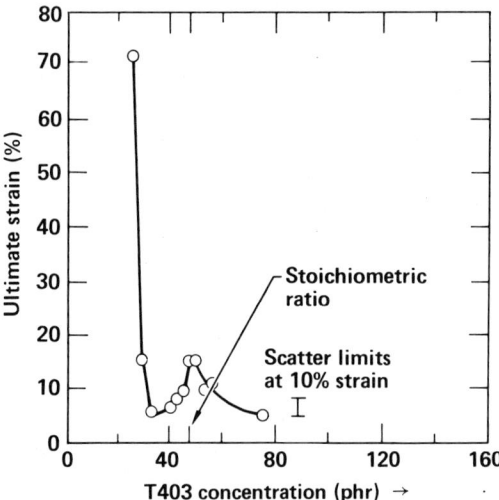

Fig. 28. The ultimate tensile strain vs. T403 concentration for DGEBA-T403 epoxies, at 23 °C

A non-random distribution of defects in the network, that produces the observed nodular morphologies in epoxies, results in weak paths in the network that favor crack propagation.

Impure starting materials will also cause network defects. We have found epoxies prepared from purified monomers with T_g's > 130 °C can exhibit excellent mechanical properties at 23 °C with tensile strengths of ~ 140 MPa and ultimate elongations of 5–8% [5].

The distribution of network segmental extensibilities controls crack initiation and propagation within the network. Embrittlement of the network can occur during plastic deformation as a result of network chain scission. There is experimental evidence that network chain scission does occur during the deformation of DGEBA-T403 epoxies. These epoxies have been strained up to 30% at 70 °C and then the load removed and this strain completely annealed out at 100 °C. The ductility of such epoxies decreased by 50% and their T_g's were lowered by 8 K compared to reference specimens that were exposed to the same thermal history but were not stressed at 70 °C. Such deterioration in the mechanical response and T_g are consistent with network deterioration via chain scission. In more direct evidence (1) stress-FTIR studies of thin DGEBA-T403 epoxy films indicate permanent chemical changes occur after removal of the load after 10% strain at 23 °C [5] and (2) Brown and Sandreczki [130] have detected free radicals by electron paramagnetic resonance studies after ballmilling these epoxies.

Hence, epoxies which prossess networks free of defects with segments of equal extensibilities would be ideal tough glasses for composite matrices.

9 Service Environment Aging

The durability of epoxy composite matrices in service environment depends on many complex interacting phenomena. The factors that control the critical path to ultimate failure or unacceptable damage depend specifically on the particular environmental conditions. These environmental factors include service stresses, humidity, temperature and solar radiation. The combined effects of thermal history, moisture exposure, and stress have a deleterious effect on the physical and mechanical integrity of epoxies. We have studied the effect of specific combinations of moisture, heat and stress on the physical structure, failure modes, and tensile mechanical properties of TGDDM-DDS epoxies [131]. Sorbed moisture plasticizes TGDDM-DDS epoxies and lowers their tensile strengths, ultimate elongations and moduli. The fracture topographies of the initiation cavity and mirror regions of these epoxies indicate that sorbed moisture enhances the craze initiation and propagation processes. Fully cured epoxy networks with segments of equal extensibilities that are tough over a wide temperature range will exhibit good wet mechanical properties and will be less susceptible to stress-moisture aging.

One of the more extreme environmental conditions experienced by an epoxy composite matrix occurs during a supersonic dash of a fighter aircraft. The aircraft dives from high altitudes (where outer surface temperature is -20 to -55 °C) into a supersonic, low-altitude run during which aerodynamic heating raises the surface temperature to 100–150 °C, in a matter of minutes. On reduction of speed, the outer surface temperature drops extremely rapidly at rates up to about 500 °C/min thus exposing the epoxy composite to a thermal spike. Simulation of such thermal spikes has been shown to increase the amount of moisture sorbed by the epoxy or epoxy composite [132-138]. However, after a certain number of consecutive thermal spikes, the amount of moisture sorbed ceases to increase. Browning [134, 137] suggested that such increases result from microcracks caused by the moisture and temperature gradients present during the thermal spike. McKague [136] noted that damage does not occur unless the thermal-spike maximum temperature exceeds the T_g of the moist epoxy. In our studies we found that the amount of moisture sorbed by TGDDM-DDS epoxies was enhanced by about 1.6 wt.-% after exposure to a 150 °C thermal spike [131]. No significant areas of microcracking were observed in the thermally spiked epoxies. We suggested the primary mechanism by which thermally spiked epoxies sorb additional moisture can be explained in terms of moisture-induced free volume changes.

10 References

1. Morgan, R. J., Kong, F. M., Walkup, C. M.: Polymer *25*, 375 (1984)
2. Morgan, R. J. et al.: Proc. 12th National SAMPE Tech. Conf. p. 368–379 (1980)
3. Pruneda, C. O. et al.: Proc. 29th National SAMPE Symp. p. 1213–1222 (1984)
4. Rinde, J. A. et al. Composites Tech. Rev. *1* (2), 4 (1979)
5. Morgan, R. J. et al.: Development of Epoxy Matrices for Filament-Wound Graphite Structures, Proc. 30th National SAMPE Symp. p. 1209–1220, LLNL Report, UCRL-90946 (1984)
6. May, C. A.: Exploratory Development of Chemical Quality Assurance and Composition of Epoxy Formulations, Air Force Materials Laboratory Report, AFML-TR-76-112 (1976)
7. Carpenter, J. F.: Quality Control of Structural Non-Metallics, McDonnell Aircraft Report, Contract No. NOOO19-76-C-0138 (1977)
8. Trujillo, R. E., Engler, B. P.: Chemical Characterization of Composite Prepreg Resins, Part 1, Sandia Laboratory Report, SAND 78-1504 (1978)
9. Mones, E. T., Morgan, R. J., Polymer Preprints, ACS, *22*, No. 2, 249 (1981)
10. Mones, E. T. et al.: Proc. of 14th National SAMPE Tech. Conf. p. 89–100 (1982)
11. Morgan, R. J., Happe, J. A., Mones, E. T.: Proc. of 28th National SAMPE Symp. p. 596–607 (1983)
12. Wolf, C. J., Grayson, M. A.: McDonnell-Douglas Research Laboratories, Private Communication
13. Pearce, P. J., Davidson, R. G., Morris, C. E. M.: J. Appl. Polym. Sci., *26*, 2363 (1981)
14. Hagnauer, G. L. and Pearce, P. J.: in Epoxy Resin Chemistry, II, Bauer, R. S. (ed.): ACS Symp. Series, *221*, Ch. 10, (1983)
15. Scola, D. A.: Proc. of 15th National SAMPE Tech. Conf., p. 9–20 (1983)
16. Morgan, R. J., Walkup, C. M., Hoheisel, T. H.: Differential Scanning Calorimetry Studies of the Cure of Carbon Fiber-Epoxy Composite Prepregs, J. Appl. Polym. Sci., *30*, 289 (1985)
17. Happe, J. A., Morgan, R. J., Walkup, C. M.: ^{1}H, ^{19}F and ^{11}B Nuclear Magnetic Resonance Characterization of BF_3: Amine Catalysts Used in the Cure of C Fiber-Epoxy Prepregs, Polymer (In press)
18. Harris, J. A., Temin, S. C.: J. Appl. Polym. Sci. *10*, 523 (1966)
19. Landua, A. J.: Polymer Preprints, ACS, *24*, 299 (1964)
20. Kaelble, D. H.: in Resins for Aerospace ACS Symposium Series 132, Editor C. A. May, ACS, Washington, D. C., Ch. 29 (1980)
21. Cizmecioglu, M., Gupta, A.: SAMPE Quarterly, *13*, No. 2, 16 (1982)
22. Gupta, A. et al.: J. Appl. Polym. Sci., *27*, 1011 (1983)
23. May, C. A. et al.: Organic Coatings Appl. Polym. Sci. Proc. *47*, 419 (1982)
24. Lee, H. and Neville, K.: Handbook of Epoxy Resins, New York, McGraw-Hill (1967)
25. Sanjana, Z. M., Schaefer, W. H., Ray, J. R.: Polym. Eng. Sci. *21*, 474 (1981)

26. Morgan, R. J., in: The Role of the Polymeric Matrix in Processing and Structural Properties of Composite Materials, Seferis, J. and Nicolais, G., New York, Plenum Press, p. 207–214 (1983)
27. Morgan, R. J., Mones, E. T.: The Cure Reactions, Network Structure and Mechanical Response of Diaminodiphenyl Sulfone Cured Tetraglycidyl 4,4'Diaminodiphenyl Methane Epoxies, Polymer (submitted)
28. Narracott, E.: Brit. Plast., 26, 120 (1953)
29. Shechter, L., Wynstra, J., Kurkjy, R. P.: Ind. Eng. Chem., 48, 94 (1956)
30. Simons, D. M., Verbanc, J. J.: J. Polym. Sci., 44, 303 (1960)
31. Anderson, H. C., SPE Journal, 16, 1241 (1960)
32. Kakurai, T. and Noguchi, T.: J. Soc. Org. Syn. Chem. Japan, 18, 485 (1960)
33. Smith, I. T.: Polymer, 2, 95 (1961)
34. Kwei, T. K.: J. Polym. Sci., 1A, 2985 (1963)
35. Stille, J. K., Culbertson, B. M.: J. Polym. Sci., A,2, 405 (1964)
36. Lee, L. H.: J. Polym. Sci., A,3, 859 (1965)
37. Price, C. C., Carmelite, D. D.: J. Am. Chem. Soc., 88, 4039 (1966)
38. Lee, H., Neville, K.: Handbook of Epoxy Resins, McGraw-Hill, New York, 1967
39. Bauer, R. S.: J. Polym. Sci., A-1, 5, 2192 (1967)
40. Sorokin, M. F., Shode, L. G., Steinpress, A. B.: Vysokomol. Soedin., A14, 309 (1972)
41. Sorokin, M. F. et al.: Vysokomol, Soedin., A14, 2420 (1972)
42. Sidyakin, P. V.: Vysokomol. Soedin., A14, 979 (1972)
43. Tanaka, Y., Mika, T. F. in: Epoxy Resins Chemistry and Technology, May, C. A. and Tanaka, Y. (eds.), New York, Dekker Ch. 3, 1973
44. Whiting, D. A., Kline, D. E.: J. Appl. Polym. Sci., 18, 1043 (1974)
45. Dušek, K., Bleha, M., Luňák, S.: J. Polym. Sci. (Polym. Chem. Ed.), 15, 2393 (1977)
46. Schneider, N. S. et al.: Polym. Eng. Sci., 19, 304 (1979)
47. Morgan, R. J.: J. Appl. Polym. Sci., 23, 2711 (1979)
48. Bokare, U. M., Gandhi, K. S.: J. Polym. Sci. (Polym. Chem. Ed.), 18, 857 (1980)
49. Stevens, G. C.: J. Appl. Polym. Sci., 26, 4259 (1981)
50. Allen, F. J., Hunter, W. M.: J. Appl. Chem., 7, 86 (1957)
51. Fowler, G. W., Fitzpatrick, J. T.: U. S. Patent 2,426,264, 1947
52. Haynes, L. T., Heilbron, I., Jones, E. R. H., Sondheimer, F.: J. Chem. Soc., 1583 (1947)
53. Fife, H. F., Roberts, F. H.: Brit. Patent, 601,608, (1948)
54. Letsinger, R. L., Traynham, J. G., Bobko, E.: J. Am. Chem. Soc., 74, 399 (1952)
55. Jacobs, T. L., Dankner, D., Dankner, H. R.: J. Am. Chem. Soc., 80, 864 (1958)
56. Burness, D. M.: J. Org. Chem., 29, 1862 (1964)
57. Bell, J. P., McCarvill, W. T.: J. Appl. Polym. Sci., 18, 2243, (1974)
58. Keenan, M. A., Smith, D. A.: J. Appl. Polym. Sci., 11, 1009, (1967)
59. Paterson-Jones, J. C., Smith, D. A.: J. Appl. Polym. Sci., 12, 1601 (1968)
60. Bishop, D. P., Smith, D. A.: J. Appl. Polym. Sci., 14, 205 (1970)
61. Leisegang, E. C., Stephen, A. M., Paterson-Jones, J. C.: J. Appl. Polym. Sci., 14, 1961 (1970)
62. Paterson-Jones, J. C.: J. Appl. Polym. Sci., 19, 1539 (1975)
63. Butler, G. B.: Polymer Preprints, ACS, 8, 35 (1967)
64. Moacanin, J. et al.: Org. Coat. and Appl. Polym. Sci. Preprints, ACS, 47, 587 (1982)
65. General Dynamics Report F33615-80-C-5021 (1980)
66. Bell, J. P.: J. Polym. Sci. A-2, 417 (1970)
67. Mijovic, J., Kim, J., Slaby, J.: J. Appl. Polym. Sci., 29, 1449, (1984)
68. Grayson, M. A., Wolf, C. J.: J. Polym. Sci. (Polymer Physics Ed.) (In press)
69. Wolf, C. J.: McDonnell-Douglas Research Laboratories, Private Communication
70. David, C., in Comprehensive Chemical Kinetics, Vol. 14, Degradation of Polymers, Bamford, C. H. and Typer, C. F. H. (eds.), Amsterdam, Elsevier, p. 1. (1975)
71. Morgan, R. J., O'Neal, J. E.: J. Macromol. Sci. Phys., B15 (1), 139 (1978)
72. Morgan, R. J., O'Neal, J. E., Miller, D. B.: J. Mater. Sci. 14, 109 (1979)
73. Morgan, R. J., O'Neal, J. E.: Polym. Plast. Technol., 10 (1), 49 (1978)
74. Kong, E. S. et al.: Polym. Eng. Sci., 21, 943 (1981)
75. Findley, W. N., Reed, R. M.: Polym. Eng. Sci. 27, 837 (1977)
76. Barton, J. M.: Polymer 20, 1018 (1979)
77. Gordon, G. A., Ravve, A.: Polym. Eng. Sci. 20, 70 (1980)
78. Diamant, Y., Marom, G., Broutman, L. J.: J. Appl. Polym. Sci., 26, 3015 (1981)

79. Carswell, T. S.: Phenolplasts, New York, Interscience, (1947)
80. Rochow, T. G., Rowe, F. G.: Anal. Chem., *21*, 261 (1949)
81. Spurr, R. A. et al.: Ind. Eng. Chem. *49*, 1839 (1957)
82. Morgan, R. J., O'Neal, J. E.: Polym. Eng. Sci. *18*, 1081 (1978)
83. Erath, E. H., Spurr, R. A.: J. Polym. Sci. *35*, 391 (1959)
84. Rochow, T. G.: Anal. Chem. *33*, 1810 (1961)
85. Gallacher, L., Bettelheim, F. A.: J. Polym. Sci. *59*, 697 (1962)
86. Erath, E. H., Robinson, J.: J. Polym. Sci. Part C, *3*, 65 (1963)
87. Wohnsiedler, H. P.: J. Polym. Sci., Part C, *3*, 77 (1963)
88. Lewis, A. F.: SPE Trans, *3*, 201 (1963)
89. Lewis, A. F., Ramsey, W. B.: Adhes. Age, *9*, 20 (1966)
90. Solomon, D. H., Loft, B. C., Swift, J. D.: J. Appl. Polym. Sci., *11*, 1593 (1967)
91. Cuthrell, R. E.: J. Appl. Polym. Sci. *11*, 949 (1967)
92. Neverov, A. N. et al.: Vysokomol. Soedin, *A10*, 463 (1968)
93. Nenkov, G., Mikhailov, M.: Makromol. Chem. *129*, 137 (1969)
94. Kenyon, A. S., Nielsen, L. E.: J. Macromol. Sci. Chem. *A3* (2), 275 (1969)
95. Strecker, R. A. H.: J. Appl. Polym. Sci. *13*, 2439 (1969)
96. French, D. M., Strecker, R. A. H., Tompa, A. S.: J. Appl. Polym. Sci., *14*, 599 (1970)
97. Turner, D. T., Nelson, B. E.: J. Polym. Sci. Polym. Phys. Ed., *10*, 2461 (1972)
98. Basin, V. Ye. et al.: Polym. Sci. USSR, *14*, 2339 (1972)
99. Kessenikh, R. M., Korshunova, L. A., Petrov, A. V.: Polym. Sci., USSR, *14*, 466 (1972)
100. Bozveliev, L. G., Mihajlov, M. G.: J. Appl. Polym. Sci., *17*, 1963 (1973); J. Appl. Polym. Sci., *17*, 1973 (1973)
101. Kreibich, U. T., Schmid, R.: J. Polym. Sci. Symp. *53*, 177 (1975)
102. Morgan, R. J., O'Neal, J. E.: Polym., Preprints, ACS, *16* (2), 610 (1975)
103. Selby, K., Miller, L. E.: J. Mater. Sci. *10*, 12 (1975)
104. Karyakina, M. I. et al.: Vysokomol. Soedin, *A17*, 466 (1975)
105. Maiorova, M. V. et al.: Vysokomol. Soedin, *A17*, 471 (1975)
106. Smartsev, V. M. et al.: Vysokomol. Soedin, *A17*, 836 (1975)
107. Morgan, R. J., O'Neal, J. E.: Polym. Plast. Technol. Eng. *5* (2), 173 (1975)
108. Racich, J. L., Koutsky, J. A.: J. Appl. Polym. Sci., *20*, 2111 (1976)
109. Morgan, R. J., O'Neal, J. E.: in: Toughness and Brittleness of Plastics, Advances in Chemistry Series 154, Deanin, R. D. and Grugnola, A. M., (eds.) (Washington, D.C. Am. Chem. Soc.), Ch. 2 (1976)
110. Morgan, R. J., O'Neal, J. E.: J. Mater. Sci. *12*, 1966 (1977)
111. Morgan, R. J., O'Neal, J. E.: in: Chemistry and Properties of Crosslinked Polymers, Labana, S. S., (ed.) New York, Academic Press, p. 289–301 (1977)
112. Luettgert, K. E., Bonart, R.: Prog. Colloid. Polym. Sci. *64*, 38 (1978)
113. Dušek, K. et al.: Polymer *19*, 393 (1978)
114. Schmid, R.: Prog. Colloid. Polym. Sci. *64*, 17 (1978)
115. Mijovic, J. S., Koutsky, J. A.: Polymer *20*, 1095 (1979)
116. Mijovic, J. S., Koutsky, J. A.: J. Appl. Polym. Sci. *23*, 1037 (1979)
117. Aspbury, P. J., Wake, W. C.: Br. Polym. J. *11*, 17 (1979)
118. Matyi, R. J., Uhlmann, D. R., Koutsky, J. A.: J. Polym. Sci., Polym. Phys. Ed. *18*, 1053 (1980)
119. Mijovic, J., Tsay, L.: Polymer *22*, 902 (1981)
120. Oberlin, A. et al.: J. Polym. Sci. (Polymer Phys. Ed.) *20*, 579 (1982)
121. Bell, J. P.: Org. Coatings & Appl. Polym. Sci. *46*, 585 (1982)
122. Takahama, T., Geil, P. H.: Makromol. Chem., Rapid Commun. *3*, 389 (1982)
123. Funke, W.: Chimia, *22*, 111 (1968)
124. Funke, W., Beer, W., Seitz, U.: Prog. Colloid. Polym. Sci. *57*, 48 (1975)
125. Fellers, J. F., Lee, J. S.: SAXS Studies of Polymeric Materials Used in High Performance Composites, Lawrence Livermore National Laboratory Progress Report, P.O. 281901 August (1983)
126. Kaelbe, D. H.: in: Epoxy Resins, Chemistry and Technology, May, C. A. and Tanaka, Y., (eds.), New York, Marcel Dekker, Ch. 5 (1973)
127. Morgan, R. J., Mones, E. T., Steele, W. J.: Polymer *23*, 295 (1982)
128. Morgan, R. J., O'Neal, J. E.: J. Polym. Sci., Polymer Phys. Ed., *14*, 1053 (1976)
129. Eyring, H.: J. Chem. Phys. *4*, 283 (1936)

130. Brown, I. M., Sandreczki, T. C.: McDonnell-Douglas Research Laboratories, St. Louis, personal communication
131. Morgan, R. J., O'Neal, J. E., Fanter, D. L.: J. Materials Sci. *15*, 751 (1980)
132. Verette, R. M.: Temperature/humidity effects on the strength of graphite epoxy laminates, AIAA Paper No. 75-1011 (1975)
133. McKague, E. L., Jr., Halkias, J. E., Reynolds, J. D. J.: Compos. Mater. *9*, 2 (1975)
134. Browning, C. E.: Ph. D. thesis, University of Dayton, Dayton, Ohio 1976
135. Hedrick, I. G., Whiteside, J. B.: Effects of Environment on Advanced Composite Structures in AIAA Conference on aircraft composites: the emerging methodology of structural assurance, San Diego, CA, Paper No. 77-463 (1977)
136. McKague, E. L.: Chapter 5 in Proceedings of Conference on Environmental Degradation of Engineering Materials', Louthan, M. R. and McNitt, R. P., (eds.), Virginia Polytechnic Inst. Printing Dept., Blacksburg, VA p. 353 (1977)
137. Browning, C. E.: The mechanism of elevated temperature property losses in high performance structural epoxy resin matrix materials after exposure to high humidity environments, 22nd National SAMPE Symposium and Exhibition, San Diego, CA *22*, 365 (1977)
138. Advanced Composite Materials-Environmental Effects, ASTM STP 658, ed. Vinson, J. P., Philadelphia, American Society for Testing and Materials, (1978)

Editor: K. Dušek
Received February 2, 1985

Mechanics and Mechanisms of Fracture of Thermosetting Epoxy Polymers

A. J. Kinloch

Imperial College of Science and Technology;
Department of Mechanical Engineering;
Exhibition Road;
London, SW7 2BX/U.K.

Thermosetting epoxy polymers are widely employed in structural engineering applications and thus a knowledge of the mechanics and mechanisms of the fracture of such materials is of vital importance. The present Chapter discusses the fracture of epoxy polymers, concentrating on the use of a continuum fracture mechanics approach for elucidating the micromechanisms of crack growth and identifying pertinent failure criteria.

List of Symbols	46
1 Introduction	47
2 Fracture Mechanics	47
3 Failure Mechanisms	48
3.1 Experimental Techniques	48
3.2 Types of Crack Growth	49
3.3 Effect of Composition	51
3.4 Effect of Test Rate and Temperature	52
3.5 Crack Tip Micromechanisms	57
4 Crack-Tip-Blunting	60
4.1 Crack Opening Displacement	60
4.2 Relation between Fracture and Yield Behaviour	61
4.3 Failure Criterion	63
5 Concluding Remarks	66
6 References	66

List of Symbols

a	crack length
a_T	time-temperature shift factor
b	specimen thickness
b_n	specimen thickness in plane of crack
c	critical distances ahead of crack
e_y	yield strain
k	geometry constant
l_m	length of moment arm
m	constant
r	distance (polar coordinate)
t_f	time-to-failure
v_f	volume fraction
w	specimen width
$\dot{y}$	rate of extension
C^*	constant
E	Young's modulus
ΔE	activation energy
G_{Ic}	mode I fracture energy
G_{Ics}	minimum value of G_{Ic}
K_{Ic}	mode I stress-intensity factor at the onset of crack growth
K_{Ics}	"sharp crack" value of K_{Ic}
P_c	applied load at onset of crack growth
R	molar gas constant
T	temperature
T_g	glass-transition temperature
δ_{tc}	crack-opening displacement at crack growth
ν	Poisson's ratio
ϱ	crack tip at crack growth
σ	stress
σ_0	applied stress
σ_{tc}	critical stress at a crack tip
σ_{yt}	uniaxial tensile yield stress
BDMA	benzyldimethylamine
CTBN	carboxyl-terminated butadiene-acrylonitrile rubber
DETA	diethylenetriamine
DDM	diphenyldiaminomethane
DGEBA	diglycidyl ether of bisphenol A
DMP	tris(dimethylaminomethylphenol)
EDA	ethylenediamine
HDA	hexamethylenediamine
HAPA	hexahydrophthalic anhydride
MNA	methylnadic anhydride
MPD	m-phenylenediamine
PA	phthalic anhydride

PIP   piperidine
TEPA   tetraethylenepentamine
TETA   triethylenetetramine

1 Introduction

The use of adhesives and fibre-composites in structural engineering applications has increased markedly in the last decade and this dramatic growth rate shows every sign of continuing in the future. Epoxy resins are widely employed as the basis for adhesive compositions and as the matrix material for glass-, polyamide- and carbon fibre composites. Thus, a knowledge of the mechanisms and mechanics of fracture of epoxy polymers is of vital importance if the strengths of components incorporating these materials are to be predicted and if improved resin systems are to be developed.

Most of the recent advances in the understanding of the fracture behaviour of epoxy polymers has been through the application of fracture mechanics [1,2] and the present Chapter is therefore concerned with the study of the mechanisms and mechanics of crack growth in crosslinked epoxies using fracture mechanics.

2 Fracture Mechanics

The reader is referred elsewhere [1-3] for a detailed discussion on continuum fracture mechanics but a very brief description will be presented here for those unfamiliar with the subject.

It may be shown that the theoretical stress to cause cleavage fracture in a brittle solid is of the order of $E/10$; that is, one tenth of the Young's modulus. The modulus of a brittle polymer, such as a typical epoxy polymer, is about 3 GPa and so the theoretical strength of such a material should be approximately 300 MPa. The measured fracture strengths of brittle polymers are well below the theoretical value (typically 40 to 100 MPa for epoxies), as is the case for many other materials. This shortfall in strengths was recognised many years ago by Griffith [4] who showed that the relatively low strength of a brittle solid could be explained by stress concentrators. This hypothesis has led to the development of a large body of experimental and theoretical work which is now termed fracture mechanics.

The basic tenet of continuum fracture mechanics is, therefore, that the strength of most real solids is governed by the presence of flaws and, since the various theories enable the manner in which the flaws propagate under stress to be analysed mathematically, the application of fracture mechanics to crack growth in polymers has received considerable attention. Two main, inter-relatable, conditions for fracture are proposed.

Firstly, the energy criterion arising from Griffith's [4], and later Orowan's [5], work which supposes that fracture may occur when sufficient energy is released (from the stress field) by growth of the crack to supply the requirements of the new fracture surfaces. The energy released comes from stored elastic or potential energy of the loading system and can, in principle, be calculated for any type of test piece. This approach, therefore, provides a measure of the energy required to extend a crack over unit area and this is termed the fracture energy and is denoted by G_{Ic}: the subscript I denoting that the crack is stressed in an tensile-opening mode (the most critical and important mode) and the subscript c that it refers to the fracture condition.

Secondly, Irwin [6] found that the stress field around a sharp crack in a linear elastic material could be uniquely defined by a parameter named the stress-intensity factor, K_I, and stated that fracture occurs when the value of K_I exceeds some critical value, K_{Ic}. Thus, K_I is a stress field parameter independent of the material whereas K_{Ic}, often referred to as the fracture toughness, is a measure of a material property. Again the subscript I is used to denote the tensile-opening mode.

A basic aim of fracture mechanics is to identify fracture criteria such as G_{Ic} and K_{Ic} and others discussed later which are independent of the geometry of the cracked body and, ideally, even of the test conditions, e.g. test temperature. Thus fracture mechanics parameters should greatly assist in developing a more fundamental understanding of the fracture process and should be of considerable use in the practical areas of material selection and development, engineering design and life prediction.

3 Failure Mechanisms

3.1 Experimental Techniques

Various test geometries may be used to determine values of the fracture energy, G_{Ic}, and stress-intensity factor, K_{Ic}, at the onset of crack growth and the more common ones are illustrated in Fig. 1.

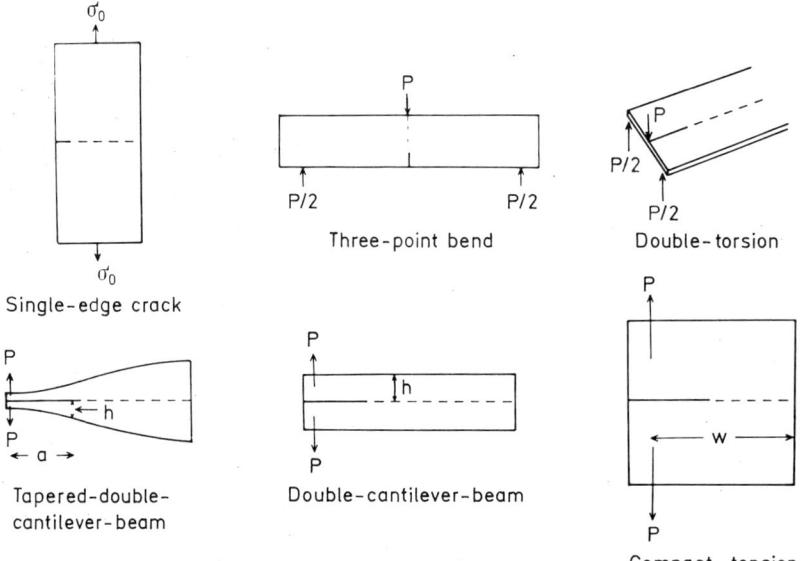

Fig. 1. Sketches of various fracture mechanics specimens [1]

Now the measured value of K_{Ic} or G_{Ic} is dependent upon the sharpness of the crack employed and an essential requirement of fracture mechanics testing is that a "naturally" sharp crack is used for the experimental determination of K_{Ic} or G_{Ic}. If a relatively blunt crack is used, an optimistically high value of K_{Ic} or G_{Ic} may be recorded. Suitable techniques include slow, controlled pressure applied to a razor blade, possibly having pre-cooled the specimen in liquid nitrogen and fatigue crack growth.

In the study of crack growth in epoxy polymers the double-torsion and compact-tension specimens have been the most widely used. The values of K_{Ic} may be obtained from [1,2)]

Double torsion:

$$K_{Ic} = P_c l_m \left[\frac{1}{lb^3 b_n (1-v) k} \right]^{1/2} \tag{1}$$

Compact tension:

$$K_{Ic} = \frac{P_c}{bw^{1/2}} \left[29.6 \left(\frac{a}{w}\right)^{1/2} - 185.5 \left(\frac{a}{w}\right)^{3/2} + 655.7 \left(\frac{a}{w}\right)^{5/2} \right.$$

$$\left. - 1017 \left(\frac{a}{w}\right)^{7/2} + 638.9 \left(\frac{a}{w}\right)^{9/2} \right] \tag{2}$$

where: P_c = load at onset of crack growth
l_m = length of moment arm
b = sheet thickness
b_n = sheet thickness in plane of crack, i.e. for grooved specimen
k = geometry constant
v = Poisson's ratio
w = width of specimen (see Figure 1)
a = crack length

In the case of linear-elastic-fracture-mechanics, and nearly all epoxy polymers obey the requirements for LEFM to be employed, a simple relationship exists between K_{Ic} and G_{Ic}

$$G_{Ic} = \frac{K_{Ic}^2}{E} \quad \text{for plane stress} \tag{3}$$

$$G_{Ic} = \frac{K_{Ic}^2 (1 - v^2)}{E} \quad \text{for plane strain} \tag{4}$$

where E is Young's modulus.

3.2 Types of Crack Growth

It has been shown [1,7,8] that, depending upon the composition of the material and upon the test conditions, three distinct types of crack propagation may be observed in thermosetting epoxy polymers. These are:

i) Stable, brittle propagation (termed type C) — here the crack grows in a steady controlled manner with the rate of crack propagation being dependent upon the crosshead speed of the testing machine. The fracture surfaces are relatively featureless and, since the cracks grow at low levels of K_{Ic} (or G_{Ic}), stable brittle propagation may be thought of as a classic example of brittle fracture. The exact shape of the associated load versus displacement curve is somewhat dependent upon the detailed specimen geometry but the example shown in Fig. 2a for a double-torsion specimen illustrates the main features of steady crack growth.

ii) Unstable brittle propagation (type B) — this type of crack growth is still essentially brittle in nature but the crack propagates intermittently in a stick/slip manner. This type of crack growth has a significant effect on the associated load versus displacement curve which now has a characteristic sawtooth shape (Fig. 2b). The values of P_i and P_a correspond to loads at crack initiation and arrest respectively. The initiation and arrest values of K_{Ic} and G_{Ic} may be determined from P_i and P_a through Eqs. (1) to (4).

iii) Stable ductile propagation (type A) — again the crack grows steadily in a controlled manner and for the double-torsion specimen the load for crack propagation is constant, as shown in Fig. 2a. However, unlike stable brittle propagation discussed

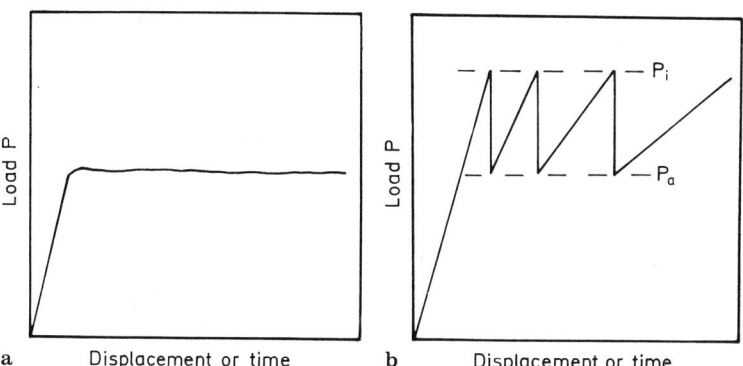

Fig. 2a and b. Schematic load versus displacement curves for crack propagation in epoxy polymers obtained using a double torsion specimen **a** Stable continuous propagation, **b** Unstable stick/slip propagation

in i), a relatively high value of K_{Ic} and G_{Ic} is now required and the fracture surfaces are far rougher and torn in appearance, indicating a more ductile fracture process.

3.3 Effects of Composition

There are a variety of structural material variables which affect the stability of crack propagation and the values of K_{Ic} and G_{Ic} at which it occurs. The variables include the type of epoxy resin and curing agent used, the amount of curing agent, and the temperature and time of cure. This is evident from Table 1 which shows values of the fracture energy, G_{Ic}, for a variety of different simple epoxy polymers. The values of G_{Ic} quoted are initiation values and all measurements were conducted at room temperature. However, it is difficult to make generalisations from the data since they have been obtained from a wide variety of resin and curing agents. Nevertheless, it can be seen that the measured values of G_{Ic} are typically of the order of 100 to 300 J/m² but they can be somewhat larger. These are considerably higher than the values of G_{Ic} obtained from other thermosetting polymers, such as phenolics or polyesters, which

Table 1. G_{Ic} values measured at room temperature for Diglycidyl ether of bisphenol A (DGEBA) epoxy resins cured with various hardeners [1]

Resin	Hardener (phr)[a]	G_{Ic} (J/m²)
Epikote 828	10 DETA	172
	95 MNA + 0.5 BDMA	154
	27 DDM	340
	14.6 MPD	110
	4 DMP	180
CT200	13 PA	220
DER 332	5 PIP	121
	various amounts TEPA	52–227
	various amounts HHPA	158–262
MY 750	8.3 EDA	329
	12.2 TDA	489
	16.1 HDA	575
	11.5 DETA	130
	11.0 TETA	141
	15.0 TEPA	136

[a] Hardeners (in parts per hundred of resin (phr))
DETA = diethylenetriamine PA = phthalic anhydride
TETA = triethylenetriamine EDA = ethylenediamine
TEPA = tetraethylenetriamine TDA = tetramethylenediamine
MNA = methylnadic anhydride HDA = hexamethylenediamine
BDMA = benzyldimethylamine PIP = piperidine
DDM = diphenyldiaminomethane HUPA = hexahydrophthalic anhydride
MPD = m-phenylenediamine DMP = tris(dimethylaminomethylphenol)

are normally less than 100 J/m² [9,10]. This demonstrates that epoxy polymers are some of the toughest thermosets available and hence are used in the most critical applications, for example, as structural adhesives or matrices for high-performance composites.

The data shown in Table 1 is for simple epoxy polymers where the cured polymer basically has a single-phase microstructure. The values of G_{Ic} can be dramatically increased, without impairing other important properties such as modulus or elevated temperature performance by establishing a multiphase microstructure [1,11,12]. Two methods have been reported based upon attaining a dispersion of a particulate second-phase in the epoxy. In one method, the second phase is rubbery in character [1,8,11-16], and a typical microstructure is illustrated in Fig. 3, whilst in the other it consists of rigid brittle particles such as alumina or silica [1,17-18]. Most recently, novel hybrid-particulate composites have been prepared and examined [20-22] which contain both rubbery and rigid dispersed phases. The improvements in G_{Ic} which may be achieved, without significantly decreasing the material's modulus or glass transition temperature, T_g, are shown in Table 2.

The underlying micromechanisms responsible for the enhancement in toughness are discussed later in Sect. 3.5.

3.4 Effect of Test Rate and Temperature

The typical effect of rate on the measured value of K_{Ic} and the associated type of crack growth can clearly be seen in Fig. 4 for an epoxy polymer based upon a DGEBA resin and cured with various amounts of TETA. The values of K_{Ici} and K_{Ica} are those for crack initiation and arrest (Fig. 2b) respectively and the difference between them characterises the amount of crack jumping which has taken place. When the difference

Table 2. Properties of multiphase epoxy polymers [22]

Composition	Property			
	T_g (epoxy) (°C)	E (GPa)	G_{Ic} (kJ/m²)	K_{Ic} (MN/m$^{3/2}$)
Epoxy polymer[a]	100	2.8	0.5	1.2
Epoxy + glass particles[b]	100	3.1	1.0	1.8
Epoxy + rubber particles[c]	100	2.2	3.3	2.7
Hybrid[d]	100	2.6	4.3	3.3
Hybrid (using silane-coated glass particles)[e]	100	2.5	5.8	3.8

[a] DGEBA/PIP cured at 120 °C/16 h
[b] Mean diameter of glass particles = 50.3 μm; volume fraction, v_f = 0.10
[c] Rubber was a carboxyl-terminated butadiene-acrylonitrile copolymer. Mean diameter of rubber particles = 1.6 μm. Rubber concentration was 15 phr, giving volume fraction of 0.18
[d] Containing: v_f (glass) = 0.1; rubber = 15 phr
[e] As (4) but glass particles were coated with γ-glycidoxypropyltrimethoxysilane to improve adhesion across particle/epoxy interface
[f] Test temperature was 50 °C

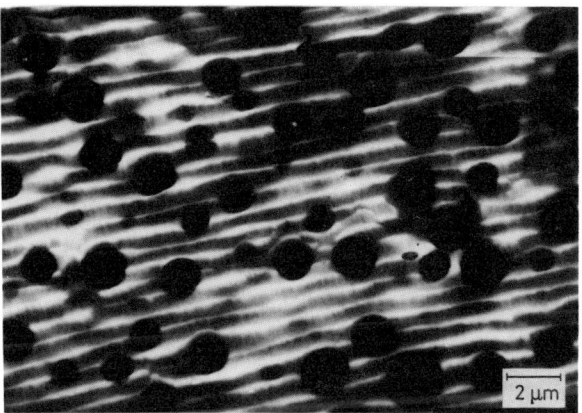

Fig. 3. Transmission electron micrograph of osmium-tetroxide stained section of a typical rubber-modified epoxy thermosetting polymer

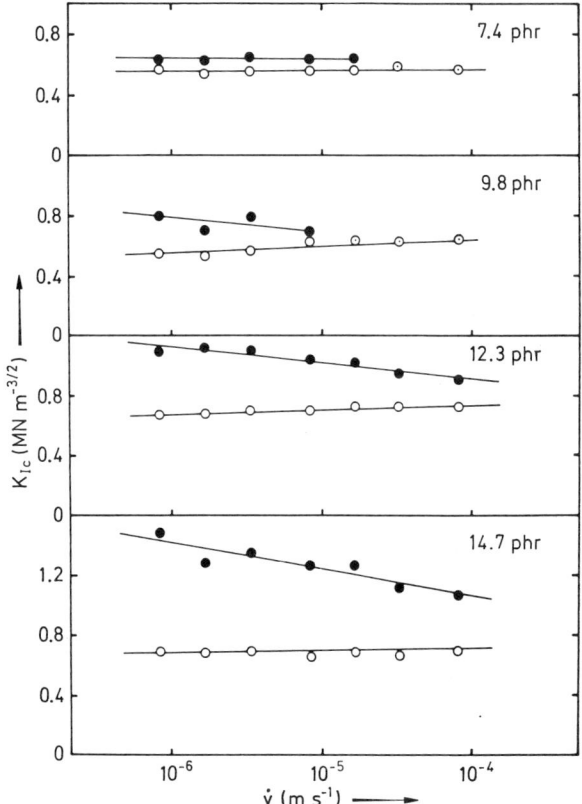

Fig. 4. Variation of K_{Ic} with crosshead speed, $\dot{y}$, for a DGEBA epoxy polymer cured with different stated phr of TETA and tested at 20 °C [9]
Brittle stable (type C) propagation: ⊙ K_{Ic},
Brittle unstable (type B) propagation ● K_{Ici}, ○ K_{Ica}

is large, propagation occurs by means of large jumps whereas when it is small, the jumps are small and when K_{Ici} is equal to K_{Ica} propagation is continuous (Fig. 2a). For all compositions of the epoxy polymer, increasing the crosshead speed (i.e. rate of testing) causes K_{Ici} to fall, with K_{Ica} remaining approximately constant, so that for some compositions the jump size decreases and eventually there is a transition to continuous propagation (type C) at high crosshead speeds.

It is possible to shed more light upon the crack-propagation process by following the effect of changing the test temperature. Fig. 5 shows the variation of K_{Ici} and K_{ica} for three different compositions of an epoxy polymer cured under identical conditions but tested at different temperatures. It is clear that in each case propagation is continuous (type C) at low temperatures but becomes unstable (type B) at higher temperatures. It is found that this behaviour is typical of other epoxy polymers and, if the test temperature range employed in Fig. 5 had been extended upwards, then ductile, stable (type A) crack growth would have been observed at the highest temperature. This is illustrated in Fig. 6.

The effects of test rate and temperature described above are obviously interrelated. This is not an unexpected observation since epoxy polymers are viscoelastic solids and so it would be predicted that reducing the rate of testing, for example, would be

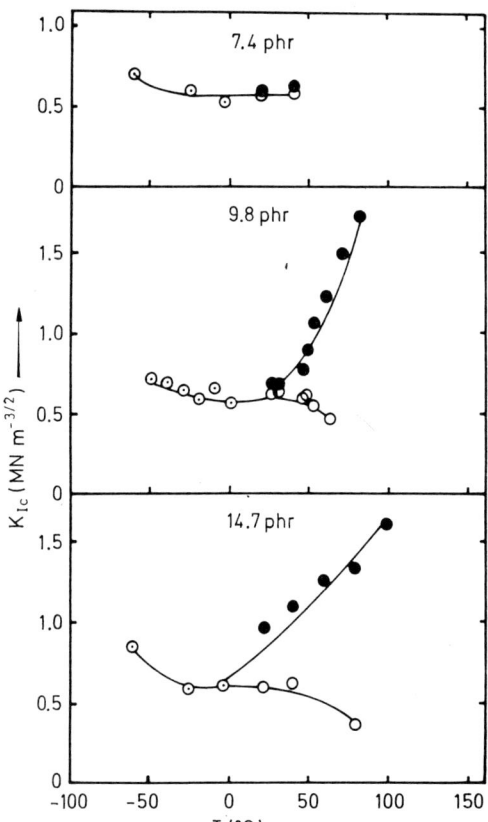

Fig. 5. Variation of K_{Ic} with temperature for the same epoxy polymer as in Fig. 4. The symbols have the same meaning and a crosshead speed of 0.5 mm/min was used [9]

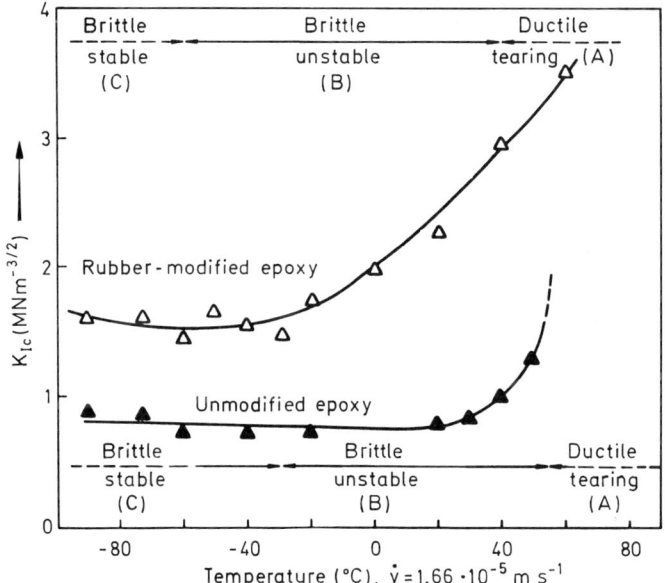

Fig. 6. Stress-intensity factor, K_{Ic}, at the onset of crack growth as a function of temperature for unmodified and rubber-modified epoxy polymers [8]

similar to increasing the test temperature — both factors promoting transitions in the crack growth behaviour from type C to type B to, eventually, type A.

This inter-relation between rate and temperature has recently been expressed in a more quantitative manner by the work of Hunston, Kinloch and co-workers [15, 23–25]. They determined the linear viscoelastic properties, such as the shear storage and shear loss modulus, for a range of unmodified and rubber-modified epoxy polymers over four to five decades of frequency and a temperature range of 80 °C. The results were then examined for the applicability of time-temperature superposition [1]. Despite the rather complex nature of the modified epoxies (see Fig. 3), it was found that the data for both the modified and rubber-modified materials superimposed well and the shift factors needed for the superposition are plotted against temperature in Fig. 7. Also shown are results from shear stress-relaxation and uniaxial compressive yield studies. The values of the shift factor, a_T, needed to time-temperature superimpose the measured properties are independent of both the type of test employed and the composition of the epoxy polymer. Thus, this indicates that whilst the absolute magnitude of the various mechanical properties depend upon composition, the shift factors do not. These shift factors may now be used to quantitatively inter-relate the values of G_{Ic} determined at different temperatures and rates; the latter parameter being characterised by the time-to-failure, t_f. The construction of a typical master curve of G_{Ic} versus reduced time-of-test, t_f/a_T, is shown in Fig. 8 for an unmodified, simple epoxy and a rubber-modified epoxy polymer. If the glass temperature, T_g, of

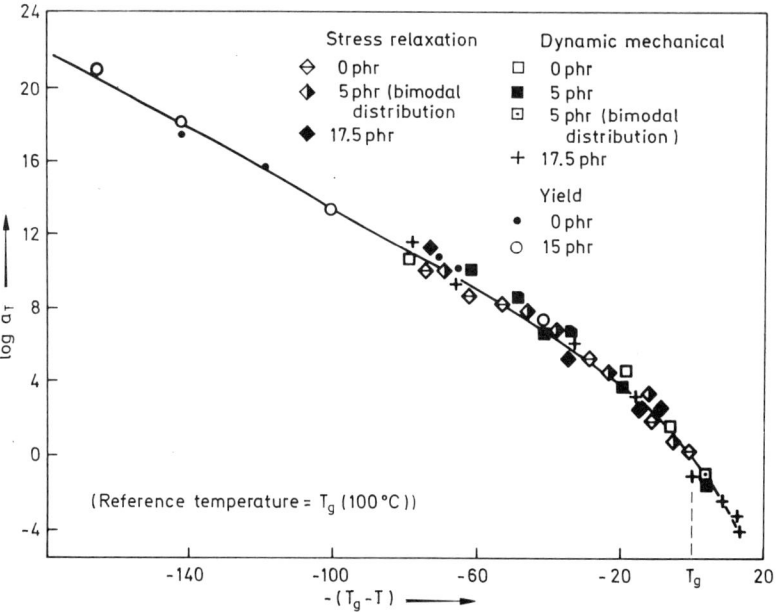

Fig. 7. Value of the shift factor, a_T, needed for superpositioning of bulk mechanical data as a function of temperature [15]. The various compositions of the rubber-modified epoxies are indicated

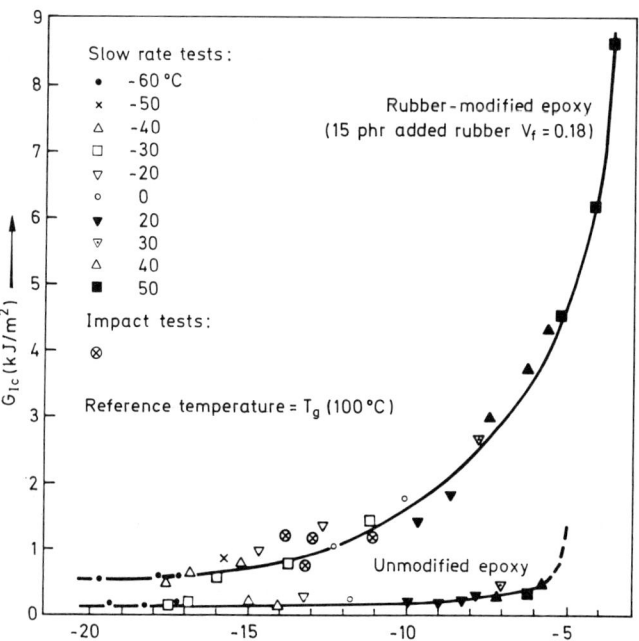

Fig. 8. Fracture energy, G_{Ic}, versus reduced time-of-failure, t_f/a_T. [15]

the epoxy is selected as the reference temperature these relationships may be modelled by

$$G_{Ic} = G_{Ics} + C^* t_f^m \, e^{-\frac{\Delta E_m}{R}\left(\frac{1}{T} - \frac{1}{T_g}\right)} \qquad (5)$$

where:

G_{Ics} = the minimum value of G_{Ic} at low temperatures and high test rates.
ΔE_m = activation energy
R = gas constant
C^*, m and E = constants which characterise the magnitude, rate dependence and temperature dependence of G_{Ic} respectively.

The power and potential of this time-temperature superposition approach is that (i) it enables different compositions to be objectively compared using an intrinsic reference point (e.g. T_g (epoxy)), and (ii) it permits the value of G_{Ic} for any given test rate/temperature combination to be predicted.

3.5 Crack Tip Micromechanisms

The results shown in Table 1 clearly reveal that the fracture energies of even unmodified, simple epoxy polymers, i.e. about 100 to 300 J/m², are at last one hundred times the energy required to break solely covalent bonds, i.e. less than 1 J/m². This demonstrates that other energy absorbing processes, such as plastic deformation, must take place at the crack tip.

Thus, even brittle crack propagation involves localised viscoelastic and plastic energy dissipative processes occurring in the vicinity of the crack tip and two such micromechanisms of major importance are shear yielding and crazing. Both involve localised, or inhomogenuous, plastic deformation of the material [1]. Shear yielding occurs essentially at constant volume and shear bands or zones develop approximately in the direction of maximum shear stress, i.e. at 45° to the maximum principal tensile stress. On the other hand, crazing occurs with an increase in volume and crazes initiate microvoids developing in a plane perpendicular to the maximum principal stress. However, such voids do not coalesce to form a true crack since they become stabilised fibrils of plastically deformed, orientated polymeric material spanning the craze. The resulting localised yielded region therefore consists of an interpenetrating system of voids and polymer fibrils and is known as a craze, and a typical craze formed in polystyrene is shown in Fig. 9.

Considering thermosetting polymers, then there is little evidence for crazing in highly-crosslinked epoxy polymers. Lilley and Holloway [27] and Morgan and O'Neal [28] have reported seeing crazes in some epoxy polymers, although the feature could have been microcracks. Van den Boogart [29] has observed crazes at the tips of cracks in undercured epoxies but could find no evidence of crazes in fully-cured polymers. Instead, he observed shear bands and this is in agreement with more recent work of Kinloch et al. [8,22]. They used scanning electron and replica transmission electron microscopy to study the fracture surfaces of a number of epoxy polymers but, whilst they found considerable evidence of plastic shear deformations, they found no fractographic evidence of crazing: fracture surfaces resulting from crazing

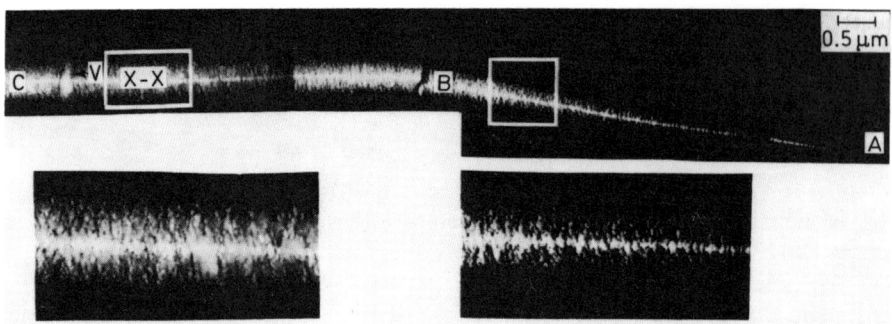

Fig. 9. Composite transmission electron micrograph of a typical craze formed in polystyrene [26]

mechanisms are very distinctive [1, 26, 30–32]. Finally, this micromechanism of plastic shear-yielding in the case of the highly crosslinked epoxies is in agreement with the theoretical work of Kramer et al. [34, 35]. Their studies demonstrated that as the crosslink density increases a transition from crazing to shear yielding results.

Turning to the multiphase thermosetting epoxy polymers (Table 2), then for the rubber-modified materials the greater fracture resistance arises from a greater extent of energy dissipating deformations occurring in the material in the vicinity of the crack tip [1, 8, 35–38]. The deformation processes are:

i) cavitation in the rubber, or at the particle/matrix interface (note, *not* in the matrix as for crazing), and

ii) multiple, but localised shear-yielding in the matrix initiated by the stress concentrations associated with the rubbery particles.

The cavitation process involves the initiation and growth of voids in the rubbery particles, which dissipates energy and gives rise to the stress-whitening that often accompanies crack growth, especially at high temperatures. This cavitation of the rubber particles is strikingly apparent in Fig. 10. However, localised plastic shear yielding is the main source of energy dissipation and increased toughness. It occurs to a far greater extent in the epoxy matrix of the rubber-modified polymers, compared to the simple epoxy polymer, due to interactions between the stress field ahead of the

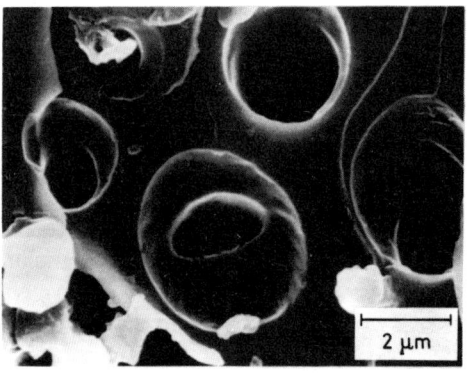

Fig. 10. Scanning electron micrograph of fracture surface of rubber-modified epoxy showing cavitated rubber particles

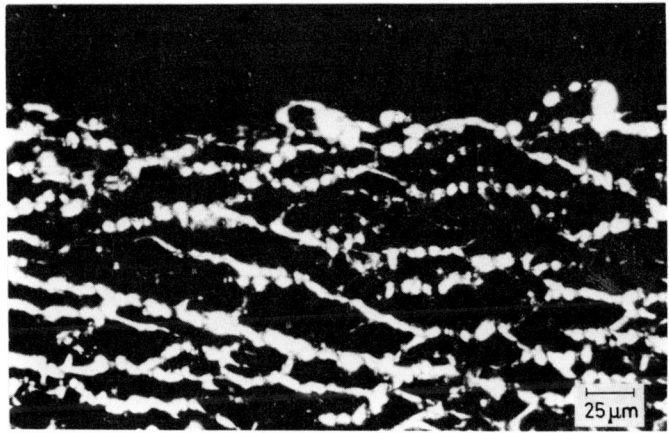

Fig. 11. Transmission optical micrograph, taken using cross polarizers, of fracture region of a rubber-modified epoxy showing shear-yield bands [37]

crack and the rubber particles. In Fig. 11a direct evidence for such shear yielding may be seen. This optical micrograph was obtained from studies on thin sections viewed between cross polars [37]. The localised shear bands are birefringent and the many hundreds of such plastic deformation regions, initiated by the stress concentrations associated with each particle, are clearly visible. In accord with previous comments it should be noted that they grow at an angle of about 45° to the principal tensile stress, i.e. in the direction of the maximum shear stress.

Considering the glass-filled and hybrid-particulate epoxy-polymers (Table 2), the latter containing both rubbery and rigid glass particles, then the presence of the glass particles enables another micromechanism to operate [22]. This micromechanism is crack pinning and basically it involves the cracks being impeded by the rigid, impenetrable, well-bonded particles. This arises since when a crack meets an array of such obstacles it becomes pinned and tends to bow out between the particles forming secondary cracks. This is illustrated schematically in Fig. 12 and by direct observation using optical microscopy in Fig. 13. Thus, new fracture surface is formed and

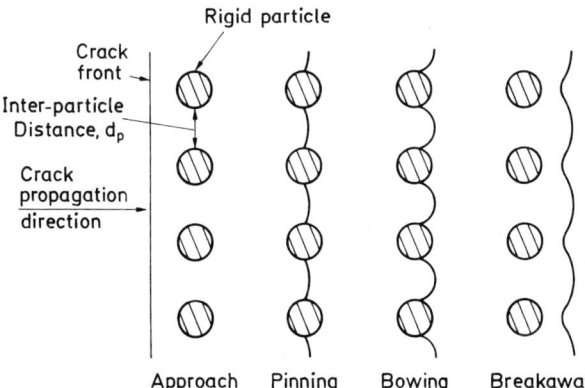

Fig. 12. Schematic representation of the crack pinning mechanism [39]

Fig. 13. Optical micrograph of fracture surface of glass-filled epoxy polymers showing the crack front pinned between glass particles [22] (Arrow indicates direction of crack growth)

the length of the crack front is increased due to its change of shape between the pinning positions. Now energy is not only required to create the new fracture surface but, by analogy with the theory of dislocations, energy must also be supplied to the newly formed non-linear crack front, which is assumed to possess a line energy. This latter factor particularly is suggested to lead to the enhanced crack resistance often observed when impenetrable particles are well-bonded into a brittle matrix. Since low adhesion is considered [40, 41] to impair the efficiency of the crack-pinning mechanism, improving the particle/matrix adhesion may result in improved toughness (see Table 2), especially when the crack-tip stresses are relatively high.

4 Crack-Tip Blunting

4.1 Crack Opening Displacement

Various models [1,2,42,43] have been proposed to describe the extent and shape of the localised plastic deformation zone at the crack tip. From these models one may define a parameter known as the crack opening displacement, δ_t (see Fig. 16) and the value of δ_{tc} for the onset of crack growth is given by

$$\delta_{tc} = \left(\frac{K_{Ic}}{\sigma_{yt}}\right)^2 e_y \qquad (6)$$

where σ_{yt} is the tensile yield stress and e_y is the yield strain; these parameters usually being obtained from uniaxial compression tests since epoxies tend to fracture before yielding in uniaxial tensile tests [44].

The value of δ_{tc} obviously reflects the degree of crack-tip blunting and values are shown for the rubber-modified epoxy polymer in Fig. 14. There are several noteworthy points. Firstly, the values of δ_{tc} increase with increasing temperature and decreasing rate. This suggests that the extent of localised plastic deformations at the crack tip, and the associated crack-tip blunting increases steadily as the temperature is increased or rate decreased, i.e. as the material's yield stress steadily falls. Secondly, accompanying this increase in crack-tip blunting, transitions in the type of crack growth are observed: type C to type B to type A as the crack becomes progressively blunter prior to propagation.

4.2 Relation between Fracture and Yield Behaviour

The above observations enable a qualitative relation, based upon the role of crack tip blunting, to be postulated between the fracture behaviour and the viscoelastic yield behaviour of the materials [7, 8, 44, 45]. The basis for this relationship is the idea that crack-tip blunting decreases the local crack tip stress concentration and thus higher applied loads are required to cause failure.

Considering the relations in detail, the crack opening displacement increasing with temperature and decreasing rate is obviously a clear indication that the extent of crack-tip blunting, resulting from localised plastic deformation at the crack tip, is increasing under these conditions. Furthermore, the results suggest that this rate/temperature dependence arises from the viscoelastic nature of the yield processes in these materials: the yield stress decreasing with increasing temperature and decreasing rate.

Therefore, turning to the rate/temperature dependence of the fracture behaviour (see Figs. 4, 5 and 6) at low temperatures (or very high loading rates) crack initiation will involve a minimum of crack tip yielding and so the crack is relatively sharp and the stress intensity factor low. This is also the case for crack arrest which is associated with very short deceleration times (i.e. very high crack tip strain rates). Thus, at low temperatures, both initiation and arrest involve relatively sharp cracks and similar loads, so stable crack growth type C results. As the temperature is increased (or the loading rate decreased) the yield stress declines (and the time during which plastic flow can occur increases) so more crack tip yielding is possible. This results in higher failure loads (larger K_{Ic} or G_{Ic} values). Once the crack begins to grow, however, the crack tip sharpens because the time period during which the material at the crack tip can yield is reduced. Consequently, the load required to propagate this sharp crack is less than that required to initiate crack growth. Only when the load falls to that associated with crack arrest the growth stops. This of course produces unstable crack growth (type B). Eventually, when the temperature is high enough (at a given loading rate) a ductile or tearing failure mode is achieved and stable crack growth, but now type A, is observed once again. The interdependence of rate and temperature in affecting the transition to ductile stable (type A) crack growth may be clearly seen in Fig. 14. As the test rate is increased, which results in less crack-tip blunting at a given temperature, a higher transition temperature is observed.

The arguments outlined above may also be readily employed to explain results from previous experiments where epoxy resins were: i) subjected to a constant load just below the short-term fracture load for a few days prior to testing [46, 47] or ii) exposed to high humidity or water for a short period prior to testing [48, 49]. These experiments showed that such test conditions can cause both an increase in toughness and a change from brittle stable to unstable crack growth. These changes can be explained in terms of the arguments presented above since such conditions would be expected to increase the ability of the material to undergo relatively greater deformation in the vicinity of the crack tip due to a lowering of the strain rate or plasticisation, respectively.

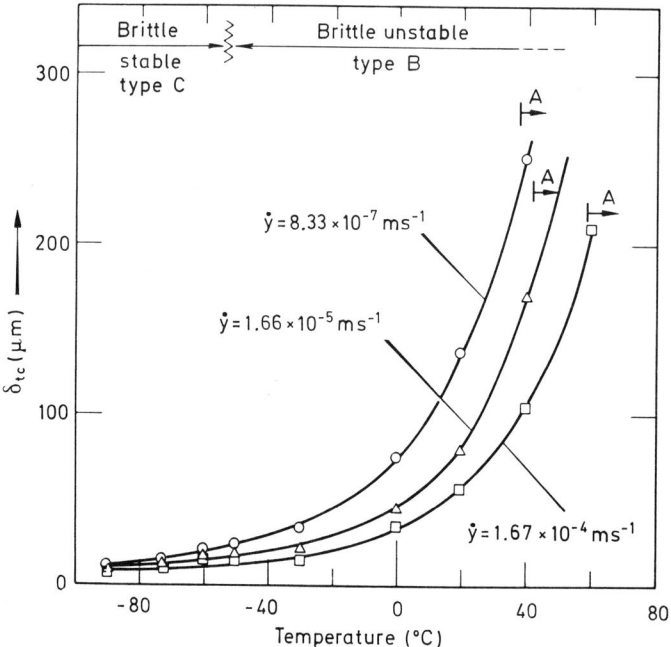

Fig. 14. Crack opening displacement, δ_{tc}, as a function of test temperature for a rubber-modified epoxy polymer [44)]
type C: brittle stable crack propagation,
type B: brittle unstable crack propagation,
type A: ductile stable crack propagation

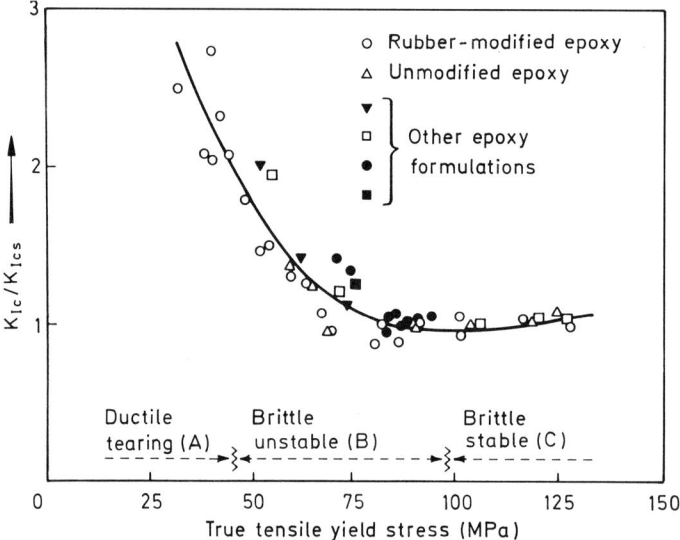

Fig. 15. Relation between K_{Ic}/K_{Ics}, true tensile yield stress and type of crack growth

Now these relations between toughness, type of crack growth and yield behaviour may be expressed in a more quantitative manner by normalising the measured values of the stress intensity factors at the onset of crack growth by K_{Ics}. The values of K_{Ics} being taken to be the stress-intensity factors measured at the lowest test temperature where the yield stress is high and the crack tip relatively sharp, as indicated by the low values of δ_{tc} shown in Fig. 14. The ratio K_{Ic}/K_{Ics} then reflects the change in toughness due to crack blunting. This ratio is shown plotted in Fig. 15 against the corresponding tensile yield stress, σ_{yt}, for various unmodified and rubber-modified epoxy polymers. A good correlation exists between the K_{Ic}/K_{Ics} ratios, yield stress and types of crack growth.

4.3 Failure Criterion

From the above observations it appears that the yield behaviour of the material in the vicinity of the crack tip controls the degree of plastic deformation that occurs locally, and hence the measured toughness and the type of crack growth. This leads to a consideration of the stress distribution around a blunt crack. It has been shown [50] that, for a crack of tip radius, ϱ, and length, a, the stress, σ_{yy}, normal to the axis of the crack at a small distance, r, ahead of the tip (Fig. 16) is given by

$$\sigma_{yy} = \frac{\sigma_0 \sqrt{a}}{\sqrt{(2r)}} \frac{1 + \varrho/r}{(1 + \varrho/2r)^{3/2}} \tag{7}$$

where σ_0 is the applied stress. Now, it has been postulated [1, 2, 44, 45, 51] that fracture occurs when a critical stress, σ_{tc}, is attained at a certain distance, c, ahead of the crack tip; $\sigma_{yy} = \sigma_{tc}$ and r = c and Eq. (7) becomes

$$\sigma_{tc} = \frac{\sigma_0 \sqrt{a}}{\sqrt{(2c)}} \frac{1 + \varrho/c}{(1 + \varrho/2c)^{3/2}} \tag{8}$$

The measured stress intensity factor, K_{Ic}, at the onset of crack growth is related to the applied stress, σ_0, by

$$K_{Ic} = \sigma_0 \sqrt{(\pi a)} \tag{9}$$

and the propagation of a "sharp" crack at a value of K_{Ics} may be interpreted as requiring the critical stress, σ_{tc}, to be attained at the distance, c, such that

$$K_{Ics} = \sigma_{tc} \sqrt{(2\pi c)} \tag{10}$$

Thus, by rearranging Eq. (8) so that

$$\frac{\sigma_0 \sqrt{(\pi a)}}{\sigma_{tc} \sqrt{(2\pi c)}} = \frac{(1 + \varrho/2c)^{3/2}}{1 + \varrho/c} \tag{11}$$

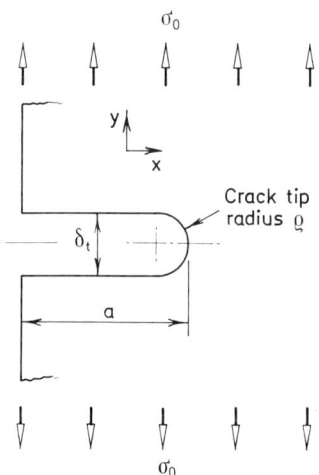

Fig. 16. Crack tip geometry

and substituting for Eq. (9) and (10) yields

$$\frac{K_{Ic}}{K_{Ics}} = \frac{(1 + \varrho/2c)^{3/2}}{1 + \varrho/c} \qquad (12)$$

Hence, the ratio K_{Ic}/K_{Ics} may be directly and quantitatively related to the crack tip radius, ϱ, at the onset of crack growth by assuming a failure criterion based upon the attainment of a critical stress acting at a certain distance ahead of the tip. A brief examination of Eq. (12) shows that it exhibits the same general trends with regard to rate and temperature dependence that were used successfully in the yield stress discussion to explain in a qualitative way the observed fracture behaviour.

Direct measurement of ϱ is difficult as it tends to be small for natural cracks in thermosetting polymers. However, Kinloch and colleagues [44, 45] have overcome this problem by measuring K_{Ic} as a function of ϱ for a series of epoxy specimens containing pre-drilled holes of large, known diameters. They showed that a relationship of the form of Eq. (12) held for these materials, as may be seen from Fig. 17. Nevertheless, there is still the problem of determining ϱ for natural cracks. Kinloch and Williams [45] have shown that extrapolation of the relationship between K_{Ic} and ϱ for artificially drilled holes to natural cracks indicates that ϱ is approximately the same as the crack-opening displacement, δ_{tc}. This is in agreement with the theoretical analysis by McMeeking [52] and experimental observations; hence ϱ can be estimated from Eq. (6).

The theory has been examined by measuring the ratio K_{Ic}/K_{Ics} as a function of $\sqrt{\varrho}$, as shown in Fig. 17. The theoretical lines have been fitted to the experimental points by choosing suitable values of the critical distance, c, which is the only fitting parameter. The agreement between theory and experiment has been found to be equally good for many different epoxy polymers cured with many different hardeners, both unmodified [44, 45, 51], rubber-modified [45] and containing glass particles [22] and even, under certain circumstances, for structural adhesive joints [53]. Values of critical stress, σ_{tc}, and distance, c, for various epoxy materials, obtained from bulk and

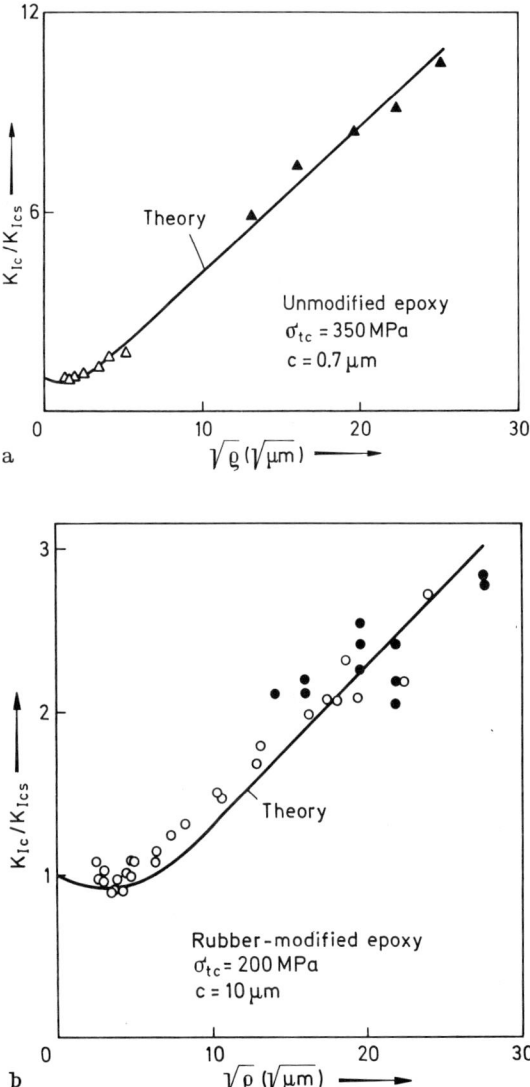

Fig. 17a and b. Variation of K_{Ic}/K_{Ics} ratio with $\sqrt{\varrho}$ [44]; **a** Unmodified, simple epoxy polymer; **b** Rubber-modified epoxy polymer; △○ values of ϱ deduced from Eq. (6); ▲● measured values of ϱ; Full curve: Theoretical relation from Eq. (12)

adhesive joint specimens are given in Table 3. The yield stress, σ_y, of each composition is also given and it can be seen that the ratio σ_{tc}/σ_y of each system is approximately the same, i.e. ∼3–5. The temperature of 25 °C is only an arbitrary reference but the constant ratio implies that the failure criterion is that a stress of the order of three- or five-times the value of σ_y must be attained at the crack tip, regardless of resin composition or yield stress. These critical stresses, σ_{tc}, may therefore be possibly interpreted as a constrained yield stress.

Table 3. Values of critical stress, σ_{tc} and critical distance, c

Composition	Configuration	σ_{tc} (MPA)	c (µm)	σ_y (25 °C) (MPa)	$\dfrac{\sigma_{tc}}{\sigma_y}$	Ref.
9.8 phr TETA	Bulk and joint	370	0.5	112	3.2	51, 53)
12.3 phr TETA	Bulk	300	1.1	91	3.3	51)
14.7 phr TETA	Bulk	270	1.6	86	3.1	51)
9.4 phr tertiary amine	Bulk and joint	360	0.4	83	4.3	45, 53)
10 phr TEPA	Bulk and joint	495	0.1	117	4.2	45, 53)
5 phr Piperidine	Bulk and joint	340	1.0	80	4.3	44, 53)
Rubber-modified epoxy	Bulk	200	10	70	2.9	44)
Hybrid-particulate epoxy	Bulk	350	1.6	75	4.7	22)

TETA: Triethylenetetramine; TEPA: Tetraethylenepentamine

To summarise, the fracture data and type of crack growth over a wide range of rates and temperatures may all be rationalised by the concept of a critical stress acting over a critical distance and this two-parameter model provides a unique failure criterion for thermosetting polymers.

5 Concluding Remarks

The micromechanisms of fracture of thermosetting polymers are dominated by localised shear-yielding at the crack tip and relationships have been established between the fracture properties and yield behaviour through a process of crack-tip blunting. However, because of the increasing use of such materials as matrix materials in high-performance composites and as high-strength structural adhesives there is still a need for more quantitative information of their fracture behaviour. This is particularly true in the area of structure/property relationships and new high-temperature tough thermosets will only be rapidly developed, and efficiently used in structural engineering applications, once such relationships have been established.

6 References

1. Kinloch, A. J. and Young, R. J.: "Fracture Behaviour of Polymers", London, Applied Science, Publ. 1983
2. Williams, J. G.: "Fracture Mechanics of Polymers", Chichester, Ellis Horwood, 1984
3. Knott, J. F.: "Fundamentals of Fracture Mechanics", London, Butterworths, 1973
4. Griffith, A. A.: Phil. Trans. Roy. Soc., *A221*, 163 (1920)
5. Orowan, E.: Repts. Prog. Phys., *12*, 185 (1948)
6. Irwin, G. R.: In "Fracture, an Advanced Treatise — Vol. 3" (ed.) Liebowitz, H., 13, New York Academic Press, 1971
7. Gledhill, R. A., Kinloch, A. J., Yamini, S. and Young, R. J.: Polymer, *19*, 574 (1978)
8. Kinloch, A. J., Shaw, S. J., Tod, D. A., and Hunston, D. L.: Polymer, *24*, 1341 (1983)
9. Young, R. J.: in "Developments in Polymer Fracture", (ed.) Andrews, E. H., 183, London, Applied Science Publ., 1979
10. Pritchard, G. and Rhoades, G. V.: Mater. Sci. Eng., *26*, 1 (1976)

11. Bucknall, C. B.: "Toughened Plastics", London, Applied Science Publ., 1977
12. Kinloch, A. J.: "Structural Adhesives: Chemistry, Microstructure and Properties", (ed.) Kinloch, A. J., London, Applied Science Publ., 1985
13. Drake, R. and Siebert, A.: SAMPE Quart., 6, 11 (1975)
14. Bascom, W. D., Cottington, R. L., Jones, R. L. and Peyser, P.: J. Appl. Polym. Sci., 19, 2545 (1975)
15. Hunston, D. L., Kinloch, A. J., Shaw, S. J. and Wang, S. S.: in "Adhesive Joints", (ed.) Mittal, K. L., 789, New York, Plenum Press, 1984
16. Chan, L. C., Gillham, J. K., Kinloch, A. J. and Shaw, S. J.: in "Rubber-Modified Thermoset Resins", (eds.) Gillham, J. K. and Riew, C. K., Washington D. C., Amer. Chem. Soc., 1984
17. Lange, F. F. and Radford, K. C.: J. Mater. Sci., 6, 1197 (1971)
18. Moloney, A. C., Kausch, H. H. and Stieger, H. R.: J. Mater. Sci., 18, 208 (1983)
19. Spandoudakis, J. and Young, R. J.: J. Mater. Sci., 19, 473 (1984)
20. Maxwell, D., Young, R. J. and Kinloch, A. J.: J. Mater. Sci., Lettrs., 3, 9, (1984)
21. Young, R. J., Maxwell, D. and Kinloch, A. J.: J. Mater. Sci. Oec. 1985 to be published
22. Kinloch, A. J., Maxwell, D. and Young, R. J.: J. Mater. Sci. Oec. 1985 to be published
23. Hunston, D. L. et al.: J. Adhesion, 13, 3 (1981)
24. Hunston, D. L. and Kinloch, A. J.: unpublished work
25. Hunston, D. L.: in "Proc. of International Adhesion Conference '84", 23, London, Plastics and Rubber Institute, 1984
26. Beahan, P., Bevis, M. and Hull, D.: Proc. R. Soc., A343, 525 (1975)
27. Lilley, J. and Holloway, D. G.: Phil. Mag., 28, 215 (1973)
28. Morgan, R. J. and O'Neal, E. J.: J. Mater. Sci., 12, 1966 (1977)
29. van den Boogart, A.: in "Physical Basis of Yield in Glassy Polymers", (ed.) Haward, R. N., London, Physics Institute, 1966
30. Doyle, M.: J. Mater. Sci., 10, 159 (1975)
31. Doyle, M.: J. Mater. Sci., 10, 300 (1975)
32. Kusy, R. R., and Turner, D. T.: Polymer, 20, 1095 (1979)
33. Donald, A. M. and Kramer, E. J.: J. Mater. Sci., 17, 1871 (1982)
34. Kramer, E. J. and Henkee, C. S.: J. Polym. Sci., Polym. Phys., 22, 721 (1984)
35. Hunston, D. L. et al.: Elast. Plast., 12, 133 (1980)
36. Kinloch, A. J. and Shaw, S. J.: J. Adhesion, 12, 59 (1981)
37. Yee, A. F. and Pearson, R. A.: "Toughening Mechanisms in Elastomer Modified Resins", NASA Contractor Report 3718, 1983
38. Pearson, R. A. and Yee, A. F.: Polym. Mater. Sci. Engng. Preprints, 49, 316 (1983)
39. Phillips, D. C. and Harris, B.: in "Polymer Engineering Composites", (ed.) Richardson, M. O. W., 45, London, Applied Science Publ., 1977
40. Green, D. J., Nicholson, P. S. and Embury, J. D.: J. Mater. Sci., 14, 1413 (1979)
41. Green, D. J., Nicholson, P. S. and Embury, J. D.: J. Mater. Sci., 14, 1657 (1979)
42. Irwin, G. R.: Appl. Mechs. Res., 3, 65 (1964)
43. Dugdale, D. S.: J. Mech. Phys. Solids, 8, 100 (1960)
44. Kinloch, A. J., Shaw, S. J. and Hunston, D. L.: Polymer, 24, 1355 (1984)
45. Kinloch, A. J. and Williams, J. G.: J. Mater. Sci., 15, 987 (1980)
46. Scott, J. M., Phillips, D. C. and Jones, M.: J. Mater. Sci., 13, 311 (1978)
47. Gledhill, R. A., Kinloch, A. J. and Shaw, S. J.: J. Mater. Sci., 14, 1769 (1979)
48. Yamini, S. and Young, R. J.: Polymer, 18, 1075 (1977)
49. Ripling, E. J., Mostovoy, S. and Bersch, C. F.: J. Adhesion, 3, 145 (1971)
50. Williams, J. G.: "Stress Analysis of Polymers", 2nd Edn, Chichester, Ellis Horwood, 1980.
51. Yamini, S. and Young, R. J.: J. Mater. Sci., 15, 1823 (1980)
52. McMeeking, R. M.: J. Mech. Phys. Solids, 25, 357 (1977)
53. Kinloch, A. J. and Shaw, S. J.: in "Developments in Adhesives — 2", (ed.) Kinloch, A. J., 82, London, Applied Science Publ., 1981

Editor: K. Dušek
Received January 9, 1985

Effect of Water on the Properties of Epoxy Matrix and Composite

A. Apicella and L. Nicolais
Materials and Production Engineering Department, University of Naples, Italy

The complex sorption behavior of the water in amine-epoxy thermosets is discussed and related to depression of the mechanical properties. The hypothesized sorption modes and the corresponding mechanisms of plasticization are discussed on the basis of experimental vapor and liquid sorption tests, differential scanning calorimetry (DSC), thermomechanical analysis (TMA) and dynamic mechanical analysis. In particular, two different types of epoxy materials have been chosen: low-performance systems of diglycidyl ether of bisphenol-A (DGEBA) cured with linear amines, and high-performance formulations based on aromatic amine-cured tetraglycidyldiamino diphenylmethane (TGDDM) which are commonly used as matrices for carbon fiber composites.

Three modes of moisture sorption are assumed: dilution of the free volume in the network, hydrogen bonding involving hydrophilic groups of the polymer and adsorption onto the surfaces of "holes" which define the excess free volume of the glassy structure and are induced hygrothermally. The DGEBA systems are described to have no appreciable sorption via hydrogen bonding, but rather most of the moisture absorbed is through dilution of the free volume and adsorption in the "holes". On the other hand, the TGDDM systems are described to absorb more moisture through significant hydrogen bonding and dilution of the free volume. High plasticization, evident as T_g depressions of 30 to 80 °C, are possible and experimentally observed especially for the stiffer TGDDM resins. Relationships derived from the free volume theory or the classical thermodynamic treatment may be used to describe the compositional dependence of the T_g in miscible blends.

Introduction	70
Solubility	71
Conclusions	76
References	76

1 Introduction

Polymer differ from other structural materials in that, at ambient temperatures, low-molecular-weight substances can easily migrate in them freely. Early studies of the sorption mechanisms and diffusion of small molecules in plastic materials arose from the desire to prepare barrier materials, mainly against gases and moisture. Recently, a number of studies of the transport properties of various substances in polymers has been performed in order to clarify the actual mechanisms of the diffusion process and to identify the factors which influence the penetrant rate of migration and the equilibrium sorption. The environmental degradation of the mechanical properties of the epoxy polymers, which are the dominant type of matrices for fiber composites, is associated to the plasticization [1-3] and micromechanical damage [4-6] induced by the sorbed moisture. Water molecules are reported to act as plasticizers or crazing agents for the epoxies, strongly influencing the properties of the material subjected to temperature, humidity and stress fatigue tests [1-10].

The chemical structure of the epoxy matrix constituent as well as processing are reported to strongly influence [11-13] the thermoset network and hence the properties and durability of the crosslinked polymer [11, 14-16]. The cure of a reactive prepolymer involves the transformation of low-molecular-weight reactive substances from liquid to rubber and solid states as a result of the formation of a polymeric network by chemical reaction of some groups in the system. Gelation and vitrification are the two macroscopic phenomena encountered during this process which strongly alter the viscoelastic behavior of the material.

Gelation is associated with a dramatic increase of the viscosity that, on a molecular level, corresponds to the increase of the molecular weight up to the incipient formation of an infinite branched molecule. Vitrification, on the other hand, corresponds to the formation of a glassy solid occurring as a consequence of the network becoming denser through further intramolecular crosslinking. The latter process may prevent further reaction by reducing the mobility of the unreacted functional groups. Therefore, even if the chemical kinetics control has been found to drive most of the polymerization process in the case of simple systems [17-19], and hence is used as a basic assumption for the statistical treatment of the cure [18, 19], sometimes, as in the case of complex commercial formulations, the cure may be controlled by physical factors, such as diffusion constrains in the glass transition region, which may possibly lead to the formation of nonhomogeneous structures that can indirectly alter the physical properties and durability of the cured resins [6]. The concept of a homogeneous infinite network is often applied to describe the morphology of all thermosetting polymers; however, the hypothesis of highly crosslinked nodules immersed in an internodular matrix of lower crosslinking density seems to better apply to some commercial formulations.

The influence of a matrix nodular structure on durability in aggressive environments, mode of failure and mechanical properties of TGDDM-DDS-based resins has been reported [20, 21]. Moreover, the incomplete cure has been recognized to reduce the durability of these materials in long-term environmental aging [11, 12]; however, the cure of high-performance high glass transition temperature matrices can be only driven to completion at high temperatures where chemical degradation may occur.

In addition, heterogeneous networks are formed at these temperatures by the occurrence of highly activated, undesired secondary reactions or by diffusion constraints.

The fact that properties differences between desiccated and soaked samples exist, supports the concept that physical modifications are introduced into the polymer network by the moisture-temperature aging. Although the exact nature of this change has not been completely identified yet, previous investigations offered explanations related to the microvoiding and crazing of the polymer [10-12] or by assuming a dual state of the water in the cured epoxies [22, 23]. In the latter case, the material would retain the penetrant even at high temperatures due to some strong mutual molecular interactions generated by the polar nature of the epoxy-water system which have been identified in several spectroscopic studies [22-24]. In systems where such interactions are present, the depression of the glass transition is greater than expected. Some studies [25], however, report that the effect of water can be related to the normal behavior theoretically predicted for the compositional dependence of the glass transition in miscible blends [26, 27]. The approach based exclusively on the free volume [28] concept often resulted in contradictory results [1, 2, 29-31] probably due to the values of T_g chosen for water or the water solubility in epoxies. However, even when correct water solubilities and water thermal parameters were used [11, 32], anomalous plasticization that could be explained only by considering strong mutual interactions between the dissolved molecules and segments of the polymer, was observed by examining systems of different composition and extent of cure [32].

2 Solubility

It has already been explained that water molecules have the tendency to form clusters and crazes and to plasticize the epoxy matrices and that they easily diffuse in the polymer [33-35]. The diffusion of water in glassy polymers that interact with the penetrant molecules is characterized by various mechanisms of sorption. The penetrant population can be divided into molecules forming an ordinary polymer-diluent solution and those adsorbed on hydrophilic sites or on the surface of excess free volume elements frozen in the glassy structure. Due to the multiplicity of the polymer-diluent interactions, the overall penetrant uptake cannot according to our opinion be used to get direct information on the degree of plasticization. Plasticization is a phenomenon associated with the depression of the glass transition temperature and mechanical properties which is induced by moisture or solvent sorption. The same amount of sorbed water may differently depress the glass transition temperature of systems with different thermal expansion coefficients or characterized by different hydrogen bonding capacity or by nodular structures which can be easily crazed in presence of the penetrant. When a polymeric material is exposed to a gas or vapor, the generally accepted mechanism for the transport of the penetrant is an activated solution-diffusion process. The molecules dissolved on the surface of the polymer diffuse through the bulk by a series of activated steps. It is clear that both the solubility and diffusivity are involved in the process and that the polymer's molecular and morphological features will influence the penetrant transport behavior. The elucidation of the mechanisms of the cure of thermosetting epoxy prepolymers may give useful insights into the nature of the polymere-diluent interactions used in the theoretical modelling of the sorption behavior.

Amines are usually used as curing agents for the epoxies: tertiary amines catalytically promote the epoxide homopolymerization, while primary and secondary amines predominantly combine with the epoxide. The nucleophilic addition of primary amines results in the formation of highly hydrophilic amino alcohols. Tertiary amines promote homopolymerization of the epoxide which leads to the formation of ether bonds. In some cases, aromatic tertiary amines may be present in the structure of some epoxies such as TGDDM and may promote etherification. This undesired reaction may compete with the addition of the secondary amine leaving unreacted amino groups able to strongly interact with the water molecules. A second common class of hardeners used for the epoxies are the organic acids, or most commonly their anhydrides. The type of reaction generates a number of hydrophilic sites (ester groups) capable of hydrogen bonding with water molecules. Generally, the nature of the polar groups depends on the specific type of hardener or additive used as well as on the processing characteristics.

Taking into account the modes in which the water can be sorbed in the resin, different models should be considered to describe the overall process. First, the ordinary dissolution of a substance in the polymer may be described by the Flory-Huggins theory which treats the random mixing of an unoriented polymer and a solvent by using the liquid lattice approach. If as is the penetrant external activity, v_p the polymer volume fraction and χ_s the solvent-polymer interaction parameter, the relationship relating these variables in the case of polymer of infinite molecular weight is as follows:

$$\ln a_s = \ln (1 - v_p) + v_p + \chi_s v_p^2 \tag{1}$$

When the solvent concentration is very small, as in the case of gas or low-activity vapor sorption, Eq. (1) becomes the limiting Henry's law and a linear sorption behavior is expected.

It should be noted that, even if several authors [7, 37] have collected a series of equilibrium sorption data for commonly used epoxy matrices exposed to humid environments that can be adequately described by Eq. (1) containing a concentration independent parameter, but often two terms of the phenomenological power series of water activity are used.

The Flory-Huggins theory, however, is not adequate for polymers containing strongly polar groups, or for the glassy state (characterized by an excess of the "free volume elements") or for polar penetrant at high uptakes (where the positive deviation from the Henry's law were interpreted [36, 38] by clustering of the penetrant). In particular, the tendency of water molecules to form clusters has been supposed to cause, especially at high temperatures, an irreversible microvoiding in the epoxies [36]. At lower temperatures the translational freedom of the water molecules is strongly hindered by the stiffness of the polymer chain segments, and the formation of clusters is kinetically unfavored and is usually not observed in short-time experiments [36]. The fact that, as will be discussed in the next section, the water solubility is increased at low temperatures due to the exothermic character of the sorption process [35] leads to the conclusion that an increase in temperature during the environmental exposure of epoxy-based materials (thermal spikes) will leave an excess of water molecules dissolved in the bulk material. Therefore, a temperature increase causes either an over-saturation of the material and a faster formation of water clusters and additional microvoiding. Hygrothermal cycling in the presence of sorbed water has been reported to induce a progressive increase of the apparent water uptake in very stiff and high T_g epoxy systems [10, 39]. Often, the first sorption isotherm in "as cast" samples is not reproducible while the second and subsequent resorptions show a reproducible behavior [32, 35]. The difference in the sorption behavior under the same external conditions has been associated with a progressively greater damage developed in inhomogeneous materials containing an increasing amount of moisture. The most important consequence of the inhomogeneous morphology from the point of view of the hygrothermal fatigue would be the difference in the rate of water diffusion and, hence, the differential swelling stresses developed between regions of different water content.

The adsorption in the "holes" generated during the vitrification process or hygrothermally induced gives rise to deviations from the normal behavior, evident as negative deviations from Henry's law. Sorption of gases and vapors is described by superposition of Henry's and Languimir's isotherms [41–43]. Similarly, the saturation of the hydrogen bonding capacity can be modelled by using a Languimir expression containing capacity and affinity constants [32] which is different from those obtained for the adsorption in "holes".

The results of the vapor sorption experiments [32, 34] have been utilized by the present authors to investigate the mechanisms of moisture sorption. The sorption curves for the TGDDM-DDS systems equilibrated at low external water activity exhibited a region of negative curvature located at about 1% uptake, which was attributed to the saturation of the hydrophilic sites. This phenomenon was observed at higher activities in epoxies containing a higher concentration of the amine hardener [32]. The greater tendency to sorption due to hydrogen bonding in systems richer in amino groups has been explained by the greater number of secondary amino groups rather than by the presence of other potential bonding sites, whose overall concentration was found unaffected by composition [11]. The equilibrium sorption isotherms determined for systems, such as the DGEBA, the curing of which occurs mainly by addition of primary and secondary amino groups [18] showed an almost linear initial behavior [33]. The formation of intermolecular bonds and the rupture of intra-chain hydrogen bonds can greatly depress the glass transition temperature. Therefore, a greater T_g depression upon equilibration with water may be expected in the stiffer and denser TGDDM systems made at a higher amino hardener concentration. Differential scanning calorimetry tests performed on wet and desiccated TGDDM-DDS samples showed a sensibly greater plasticization for the system of higher DDS content even though the same amount of water was sorbed by systems of a lower hardener concentration [2, 32]. An interesting effect of the composition on plasticization also has been described for

Table 1. Volume expansion change at T_g density and glass transition temperature of amine hardened TGDDM-based resins

DDS, PHR	$\alpha \cdot 10^3$, C^{-1}	ϱ, g/cm^3	T_g, C
20	0.63	1.262	150
30	1.08	1.273	175
50	2.95	1.284	200

these systems [32]; the samples of intermediate DDS concentration exhibited a lower plasticization than the systems of higher and lower amine content. This anomalous behavior was explained by free volume considerations [32]. As indicated before, the plasticization of the network structure depends on the dilution process as well as on intermolecular bond formation. The former is governed by the diluent volume fraction, V_d, and by the incremental change of the thermal expansion coefficients at the glass transition, α_d and α_p, through the well-known expression [28]:

$$T_{gw} = T_{gp} \frac{V_p \alpha_p}{V_p \alpha_p + V_d \alpha_d} + T_{gd} \frac{V_d \alpha_d}{V_p \alpha_p + V_d \alpha_d} \qquad (2)$$

where the subscripts p and d refer to the polymer and the diluent, respectively. According to this expression, the higher the incremental change in the polymer thermal expansion coefficient at T_g, the weaker the influence of the diluent volume fraction. This is due to the fact that the glass transition of a wet polymer is calculated as the mean value of the characteristic temperatures of the constituents weighted by the product of the volume fraction and thermal expansion coefficient. By comparing the values of the thermal expansion coefficients and densities reported in some of our previous publications [12, 32] for TGDDM-DDS systems and summarized in Table 1, it can be seen that as the both the density and the thermal expansion coefficient increase with increasing concentration of the hardener. This result indicates that, according to Eq. (2), the dilution and plasticization process becomes less important as the DDS concentration increase. In TGDDM-DDS systems at a low DDS concentration, plasticization occurs primarily by dilution while the influence of the hydrogen bonding on hydrophilic sites, which is operative at very low water activities [32], is less important. Conversely, the sorption and plasticization of DDS-rich materials are primarily related to the presence of the unreacted amines.

Recently, alternative theoretical expressions have been developed by using classical thermodynamic treatments to describe the compositional dependence of the glass transition temperature in miscible blends and further extended also to the epoxy-water systems [25, 27]. The studies carried out on DGEBA epoxy resins of relatively low glass transition have shown that the plasticization induced by water sorption can be described by theoretical predictions given by:

$$T_{gw} = \frac{\Delta C_{pp} X_p}{\Delta C_{pp} X_p + \Delta C_{pd} X_d} + T_{gd} \frac{C_{pd} X_d}{\Delta C_{pp} X_p + \Delta C_{pd} X_d} \qquad (3)$$

where X is the weight or mole fraction and ΔC_p is the incremental change in the specific heat at the glass transition. Actually, the value of the ΔC_p to be used should refer only to the units capable of activation [26]; however, the experimentally measured DSC value may be used in the case of epoxy systems. In fact, when highly crosslinked networks are produced, it can be assumed that all units are involved in cooperative molecular motion at T_g. The lower the value of ΔC_{pp}, the greater is the expected depression of the glass transition temperature especially at low concentrations of plasticizer. Ellis and Karasz [25] have theoretically calculated T_g depressions of 10 to 15 °K/wt% water for the less crosslinked DGEBA systems and of 25 °K/wt% water for the high T_g TGDDM-based systems by using a value of ΔC_{pd} for the water of 1.94 J °K [44] and values ranging from 0.34 to 0.53 for the epoxies. These expectations were generally confirmed by their experiments on samples equilibrated at high temperatures. In some cases, however, the calculated plasticizations were slightly greater than expected. The observed differences should be attributed principally to the experimentally measured values of the water uptake and not to the deficiency of the model. The samples equilibrated at a high temperature exhibited higher water uptakes [6,32] and showed much lower plasticization as determined by mechanical tests than the low-temperature conditioned samples. Table 2 resumes this apparently anomalous behavior of an

Table 2. Ratios of the elastic moduli and strengths of the dry and water equilibrated DGEBA-based resin at different temperature: 25 °C

Water equilibration Temperature, °C	E_w/E_d %	σ_{bw}/σ_{bd} %	Water uptake, %
2	64	77	2.58
20	69	80	2.96
50	72	86	3.22
70	83	92	4.00

amine-cured DGEBA resin. As previously discussed, sorbed moisture may be present in the polymeric network in different forms associated with an equilibrium content which is dissolved or bonded in the compact resin and with an excess moisture stored in preexisting or hygrothermally formed "holes". Assuming that only dissolved water contributes to the plasticization of the network, the higher microcavitation expected to occur at high temperatures will induce high water uptakes even though the actual water solubility is a decreasing function of temperature.

High plasticizations, reflected in the glass transition temperature depression of the order of 10 to 20 K per percent of sorbed water, are experimentally observed [6] and theoretically predicted [25,32] for commonly used epoxy systems such as those based on DGEBA and TGDDM resins. Moisture uptakes of 2—5% can easily be reached under environmental conditions [6,34] leading to a significant drop of T_g from 150—200 °C down to 60—80 °C, well below the application limits of most of the epoxy-based composites. Some tests in progress on high-performance carbon fiber composites obtained from the commercial type of TGDDM-based, preimpregnated materials equilibrated in water at 5 °C exhibited the calorimetrically measured T_g depressions of more than 100 °C. Additional dynamic mechanical studies are in progress in order to further clarify the actual mechanism of plasticization.

Previous dynamic mechanical experiments reinforced the basic assumptions of the proposed sorption and plasticization mechanisms. Low glass transition DGEBA-based epoxies and high glass transition TGDDM-based epoxies have been described [39] to exhibit transitions at high temperatures associated with the glass transition and quite broad low-temperature transitions. While the glass transition requires large-scale movements, the secondary transitions are often a combination of molecular rotation of some main chain side groups, motion of some segments of the main chain, or motion of small molecules dissolved in the polymer. Thus, changes in a polymer network structure are manifested by transitions exhibited in dynamic mechanical spectra. The broad low temperature transition indicated a wide spectrum of motions and activation energies [45]. The water conditioning of DGEBA and TGDDM resins resulted in an increase of the magnitude of the transition and a slight drop in the transition temperature of about 20–25 K, with increasing moisture content. By plotting the area under the loss compliance transition as a function of the moles of sorbed water per mole of nitrogen in the resin it has been observed that saturation of the available amino-hydrogen bonding sites in DGEBA systems occurs at a lower moisture content than in the TGDDM system [46]. Thus, the dynamic mechanical data support the assumption that the sorption by hydrogen bonding is more pronounced for the TGDDM systems than for the DGEBA system.

3 Conclusions

The analysis of vapor and liquid sorption data as well as DSC, TMA, dynamic mechanical and density experiments revealed that the dilution of free volume in the network structure and hydrogen bonding to hydrophilic sites are two possible mechanisms by which moisture is sorbed in epoxy systems. In particular, for a high glass transition temperature system, such as the TGDDM epoxy resin cured with aromatic amines, the importance of the hydrogen bonding mechanism in the sorption and plasticization increases with increasing concentration of the hardener, while the dilution mechanism becomes less significant. This is attributed to an increase in the number of hydrophilic sites such as unreacted amines and to the more crosslinked network. Although in a more dense and crosslinked network the number of hydrophilic sites is higher, the ability of the moisture to diffuse into the available free volume is lower. In low glass transition temperature and less stiff DGEBA-based epoxy systems, no appreciable sorption was found to occur by the hydrogen-bonding mechanisms, as expected for crosslinking occurring mainly by reaction of all amine groups. Therefore, most of the sorption can be attributed to the dilution of the free volume. An additional insight into the mechanism of sorption and plasticization can be obtained by using dynamical mechanical tests. Equilibrium dynamic tests indicate a significant variation in the transition of the tan δ curve. An increase in the difference in area under the loss compliance curves of dry samples and samples containing different amount of water is described by an increase in the molar ratio of moisture to nitrogen content. Specifically, the DGEBA-based systems show a saturation point where additional moisture entering the network does not induce any additional hydrogen bonding. Conversely, this saturation plateau is not found for the TGDDM-based system, which confirms the importance of the hydrogen bonding mechanism.

4 References

1. Browning, C. E.: Polym. Eng. Sci., *18*, 16 (1978)
2. Morgan, R. J., Mones, E. T., Steele, W. J.: Polymer, *20*, 315 (1982)
3. Morgan, R. J., O'Neal, J.: J. Mater. Sci., *12*, 1966 (1977)
4. Apicella, A., Nicolais, L.: Ind. Eng. Chem. Prod. Res. Dev., *20*, 138 (1981)
5. Chi-Hung, Springer, G. S.: J. Compos. Mater., *10*, 2 (1976)

6. Apicella, A., Nicolais, L.: Ind. Eng. Chem. Prod. Res. Dev., *23*, 288 (1984)
7. Loos, A. C., Springer, G. S.: J. Compos. Mater., *13*, 17 (1979)
8. Pogany, G. A.: Polymer, *17*, 690 (1976)
9. McKague, E. L., Halkias, J. E., Reinolds, J. D.: J. Compos. Mater., *9*, 2 (1975)
10. Apicella, A., Nicolais, L., Astarita, G., Drioli, E.: Polymer, *20*, 9 (1979)
11. Apicella, A., Nicolais, L., Carfagna, C., Notaristefani, D. de: C. Voto, Proceedings of the 27th National SAMPE Meeting, San Diego USA, 1982, "The Effect of the prepolymer composition on the environmental aging of epoxy based resins"
12. Apicella, A., Nicolais, L., Halpin, J. C.: Proceedings of the 28th National SAMPE Meeting, Anaheim, USA, 1983, "The role of the processing chemorheology on the environmental ageing behaviour of high performance epoxy matrices"
13. Apicella, A.: "Influence of chemorheology on the epoxy resin properties", to appear in "Development of reinforced plastics-5, Ed. Pritchard, Appl. Sci. Publishers LTD.
14. Gillham, J. K.: Polym. Eng. Sci., *19*, 676 (1979)
15. Lewis, A. F., Doyle, M. J., Gillham, J. K.: Polym. Eng. Sci., *19*, 687 (1979)
16. Mijovic, J., Tsai, L.: Polymer, *22*, 902 (1981)
17. Lunak, S., Vladyka, J., Dusek, K.: Polymer, *19*, 931 (1978)
18. Dušek, K., Bleha, M., Lunak, S.: J. Polym. Sci. Polym. Chem. Ed.. *15*, 2393 (1977)
19. Dušek, K., Ilavsky, M.: Colloid Polym. Sci., *28*, 605 (1980)
20. Schneider, N. S., Sprouse, J. F., Hagnouer, G. L., Gillham, J. H.: Polym. Eng. Sci., *19*, 304 (1979)
21. Morgan, R. J., O'Neal, J., Miller, D. B.: J. Mater. Sci., *14*, 109 (1979)
22. Moy, P., Karasz, F. E.: Polym. Eng. Sci., *20*, 315 (1980)
23. Banks, L., Ellis, B.: Polymer Buil., *1*, 377 (1979)
24. Anton, M. K., Koening, J. L., Serafini, T.: J. Polym. Sci., Polym. Phys. Edn., *19*, 1567 (1981)
25. Ellis, T. S., Karasz, F. E.: Polymer, *25*, 664 (1984)
26. Ellis, T. S., Karasz, F. E., Brinke, G. ten: J. Appl. Polym. Sci., *28*, 23 (1983)
27. Brinke, G. ten, Karasz, F. E., Ellis, T. S.: Makromolecules, *16*, 244 (1983)
28. Kelly, F. N., Bueche, F.: J. Polym. Sci., *50*, 549 (1961)
29. Peyser, P., Bascom, W. D.: J. Mater. Sci., *16*, 75 (1981)
30. Morgan, R. J., O'Neal, J. E.: Polym.-Plast. Tech. Eng., *10*, 49 (1978)
31. McKague, E. L., Reynolds, J. D., Halkais, J.: J. Appl. Polym. Sci., *22*, 1643 (1978)
32. Apicella, A., Nicolais, L., Cataldis, C. de: "Characterization of the morphological fine structure of commercial thermosetting resins through hygrothermal experiments", Advance in Polymer Science vol. 66, Kausch Ed., Springer-Verlag 1984
33. Apicella, A., Tessieri, R., Cataldis, C. de: J. Memb. Sci., *18*, 211 (1984)
34. Apicella, A., Nicolais, L., Astarita, E., Drioli, E.: Polymer, *22*, 1064 (1981)
35. Apicella, A., Nicolais, L., Astarita, G., Drioli, E.: Polym. Eng. Sci., *21*, 18 (1981)
36. Carfagna, C., Apicella, A.: J. Appl. Polym. Sci., *28*, 2881 (1983)
37. Delasi, R., Whitside, J. B.: ASTM STP 658, J. R. Vinson Ed., 1978, p. 2
38. Zimm, B. H., Lundenberg, J.: J. Chem. Wash., *60*, 425 (1956)
39. Mikols, W. J., Seferis, J. C., Apicella, A., Nicolais, L.: Polym. Composites, *3*, 118 (1982)
40. Carfagna, C., Apicella, A., Nicolais, L.: J. Appl. Polym. Sci., *27*, 105 (1982)
41. Michaels, A. S., Vieth, W. R., Barrie, J. A.: J. Appl. Phys., *24*, 1 (1963)
42. Meares, P.: Trans. Farad. Soc., *53*, 101 (1957)
43. Vieth, W. R., Howell, J. H., Hoseih, J. H.: J. Memb. Sci., *1*, 177 (1977)
44. Johari, G. P.: Phylos. Mag., *35*, 1077 (1977)
45. Keenan, J. D., Seferis, J. C., Quinlivan, J. T.: J. Appl. Polym. Sci., *24*, 2375 (1979)
46. Apicella, A., Nicolais, L., Mikols, J. K., Seferis, J. J.: "Sorption mechanisms in glassy thermosets" in "Interrelations between processing structure and properties of polymeric materials", J. C. Seferis and P. S. Theocaris (Eds.), Elsevier 1984

Editor: K. Dušek
Received: February 4, 1985

Siloxane-Modified Epoxy Resins

E. M. Yorkgitis, N. S. Eiss, Jr., C. Tran, G. L. Wilkes, and J. E. McGrath
Departments of Chemical Engineering, Mechanical Engineering, and Chemistry and Polymer Materials and Interfaces Laboratory Virginia Polytechnic Institute and State University Blacksburg, VA 24061/USA

Epoxy resins chemically modified with functionally terminated poly(dimethyl siloxane), poly(dimethyl-co-methyltrifluoropropyl siloxane), and poly(dimethyl-co-diphenyl siloxane) oligomers are described in terms of their synthesis, morphology, solid-state properties, and friction and wear properties. The compatibility between the epoxy resin and the siloxane modifiers can be enhanced by increasing the percentage of methyltrifluoropropyl (TFP) siloxane or diphenyl (DP) siloxane relative to dimethyl siloxane. The compatibility of the siloxane modifier with the epoxy resin subsequently controls the size and make-up of the phase-separated elastomeric domains. Fracture toughness of the epoxy resin can be improved by modification with siloxanes containing 40% or higher TFP content or 20 and 40% DP content. Both fracture toughness and modulus changes are given a morphological basis. Friction and wear properties in both fatigue and abrasive wear tests have been found to be related to changes in modulus, fracture toughness, and morphology in ways dependent on the particular modifier. The properties and morphology of epoxy resin modified with amine-terminated and carboxyl-terminated butadiene acrylonitrile elastomers prepared under similar curing conditions have been determined and compared to those of the siloxane-modified resins.

1 Introduction 80
2 Experimental 83
   2.1 Materials 83
   2.2 Sample Designation 84
   2.3 Static Mechanical Properties 84
   2.4 Dynamic Mechanical Properties 85
   2.5 Scanning Electron Microscopy 85
   2.6 Friction and Wear 85

3 Results and Discussion 86
   3.1 Characteristics of the Siloxane-Modified Epoxy Networks 86
   3.2 Morphology and Solid-State Properties 91
   3.3 Friction and Wear Properties 102

4 Concluding Remarks 107

5 References 108

1 Introduction

Since the first epoxy resin patents were granted in the 1930's and 40's, the properties of epoxy resins, for example, excellent chemical resistance, very good adhesion, and good electrical insulation, have been utilized in many applications [1, 2, 3]. These include surface coatings, adhesives, castings, and laminates. The versatility of these cross-linked systems stems in large part from the fact that one can choose from a wide variety of resins, curing agents, and preparatory conditions and often tailor a resin to suit a particular need.

Current demands for so-called high performance materials has heightened interest in epoxy resins as structural adhesives and as matrix resins for high-strength composites. Both of these applications take advantage of the resin's outstanding strength and modulus and generally good adhesion. However, such uses also require good fracture resistance and impact strength, properties which epoxy resins do not generally exhibit.

The most common route to toughening epoxy resins has involved the incorporation of elastomeric modifiers into the final glassy matrix. A reactive liquid elastomer is mixed with the resin and curing agent at relatively low temperatures to form a homogeneous mixture. As cure proceeds (generally at elevated temperatures), the increasing molecular weight of the epoxy matrix forces the elastomeric component to separate within the crosslinked resin and form a second dispersed phase. Unlike thermoplastics modified with rubbers through blending, the ideal rubber-modified thermoset links the resin and the elastomer through covalent bonds such that, in the final state, the rubber, resin, and curing agent form a multi-phase three-dimensional glassy network. As will be discussed, the dispersed elastomeric domains act to alleviate crack propagation through various proposed mechanisms.

The principal objective of rubber modification is the improvement of fracture properties with the smallest possible decrease in modulus and strength. The CTBN and ATBN (carboxyl- and amine-terminated butadiene acrylonitrile) copolymers have done much towards reaching this end. The first publications by McGarry [4, 5] as well as B. F. Goodrich co-workers Rowe, Siebert, and Drake [6, 7] reported strong improvements in fracture surface work and fracture energy, respectively, with the addition of 5 to 15% CTBN. Particle sizes in the best of these systems range from 0.1 to 1.0 micron. Riew, Rowe, and Siebert [8] have reported 30- to 40-fold increases in fracture energy. Although they are successful as general purpose toughening agents for epoxies, CTBN and ATBN elastomers possess two drawbacks. First, their glass transition temperatures are high relative to most elastomers [9], which limits their application far below room temperature [10, 11]. Second, their highly unsaturated structure provides possible sites for reaction in oxidative and high temperature environments [12].

Siloxane elastomers present an attractive alternative to the butadiene acrylonitrile elastomers most often used for epoxy modification. Poly(dimethyl siloxanes) exhibit glass transition temperatures well below those of butadiene acrylonitrile modifiers (minimum -123 °C vs. about -40 °C) and also display very good thermal stability [13, 14]. Other favorable and potentially useful attributes include good weatherability, oxidative stability, and moisture resistance. Finally, the non-polar nature and low surface energy of poly(dimethyl siloxanes) constitute a thermodynamic driving force for them to migrate to the air-polymer interface, provided the chains are sufficiently mobile. This migration can occur with simple physical blends as well as in systems with chemically linked microphase-separated segments. During the early stages of cure of a siloxane-modified epoxy, before extensive crosslinking limits diffusion, such migration is considered possible and is believed to lead to the formation of a very hydrophobic and chemically bound surface coating [15]. Evidence suggests that this non-fugitive "slippery" surface layer enhances the friction and wear properties of the epoxy substrate [16].

This chapter is meant to be an overview of ongoing studies of polysiloxane-modified epoxy resins. Because this research area is still quite young, it is not yet possible to write a standard review article. Presented here is the current status of a collaborative effort encompassing chemistry and synthesis of the modified networks, their morphology, their mechanical properties, and their friction and wear behavior. The earliest work in the synthesis and characterization of siloxane-modified networks was done by Riffle et al. [15]. More recent research in the area of chemistry and synthesis has been carried out by Tran [17].

The extensive body of literature on ATBN- and particularly CTBN-modified epoxies provides a wide base for future work in the general area of rubber modification of epoxies. Bascom, Hunston, and coworkers [18-22] have systematically studied the fracture behavior of materials containing up to 30 weight percent CTBN and elucidated the differences between their bulk and adhesive fracture properties. Sayre, Assink, Lagasse, and Kunz [23,24] have studied in detail the phase structure and composition of CTBN-modified epoxy resins. Manzione, Gillham, and McPherson [25,26] have demonstrated how the nature of the cure, phase structure, and, ultimately, the mechanical properties can be controlled by acrylonitrile content of the elastomer and the curing conditions. Bucknall and Yoshii [27] pointed out the influence of several factors, including type and concentration of curing agent and solubility parameter and molecular weight of the modifier, on the final properties of a toughened epoxy resin.

It is not yet clear how the addition of an elastomer to an epoxy network acts to increase fracture resistance. McGarry [4] and Bucknall [28] initially proposed that these modified glassy networks absorbed mechanical energy through crazing, much like high-impact polystyrene. This view was later found to be inadequate, particularly when applied to highly crosslinked thermosets such as those used for structural laminates. Further work towards an understanding of mechanisms is described by Yee [29] and Kunz et al. [10,30].

Kinloch, Shaw, Tod, and Hunston [11] recently studied the fracture toughness behavior of CTBN-modified epoxy resins at temperatures from -93 to 60 °C and at several displacement rates. Three basic types of crack growth were defined, and the corresponding fracture surface features were identified. The authors evaluated the principal mechanisms proposed for rubber toughening — rubber tearing [10,30], crazing [4,28], and combination shear yielding and crazing [8,31] — and described a collective rubber toughening mechanism. They concluded that the primary source of energy dissipation in unmodified and CTBN-modified epoxies is yielding and plastic shear flow of the matrix. The formation of voids at the domain/matrix interface or within the domains themselves was regarded as a secondary but still important source of energy dissipation. Kinloch et al.[11] point out that the interplay between these two mechanisms ultimately determines the contribution of each. A companion paper by Kinloch, Shaw, and Hunston [32] introduced the critical values of the applied stress and the crack tip radius as a unique fracture criterion for rubber-modified epoxies. Their relationship to the physical processes active during fracture is as yet unexplored.

The compatibility of the modifier with the resin system is considered an important factor in achieving toughening [28, 33]. The solubility parameter is a good indicator of the compatibility of one substance with another and, together with molecular weight and temperature, can adequately predict the nature of the phase separation of the elastomer from the resin during cure. Through copolymerization of dimethyl siloxane with partially aromatic diphenyl (DP) siloxane or polar methyl trifluoropropyl (TFP) siloxane, one can raise the solubility parameter of the siloxane elastomer from 7.5 $(cal/cc)^{1/2}$ close to that of the epoxy resin, approximately 9.2 [6]. This is analogous to the manner in which the solubility parameters of ATBN and CTBN elastomers are controlled through acrylonitrile content [6].

The control resin network used in this study was a diglycidyl ether-based epoxy resin crosslinked with a cycloaliphatic diamine. Cooligomeric modifiers were prepared having varying percentages of TFP and DP siloxane and aminoethylpiperazine end groups. Both siloxane and ATBN and CTBN elastomers were used as epoxy modifiers, the latter two having been included to facilitate direct comparisons between modifiers in similarly prepared networks.

Ideally, rubber toughening should be accomplished without substantial sacrifices in modulus. For each modified resin, flexural and Young moduli and plane-strain fracture toughness were determined. Examination of various fracture surfaces by scanning electron microscopy showed the effects of modifier composition on the morphology of these multi-phase materials as well as the prominent features of the fracture process.

As will be discussed, incorporation of siloxane oligomers modified the elastic moduli and the fracture properties of the crosslinked epoxy network. Previous work [15] indicated that the surface of these materials was rich in siloxane, which is believed to foster a low energy surface. These characteristic properties have led to our interest in the friction and wear of siloxane-modified epoxies.

Experimental determination of the friction force considers that the work done by that force is equal to the deformation losses in the materials in the vicinity of the contact points. These losses are a combination of elastic hysteresis and plastic deformation. Adhesive forces between the materials at the contact points determine the magnitude of the tangential forces on the surface [34]. Thus modifiers to materials which decrease their surface energy are expected to reduce the friction forces. Modified epoxies were accordingly studied as a function of the amount and type of modifier as well as the normal load. Materials were studied using two experimental procedures, each of which was designed to promote a particular wear mechanism. The four major wear mechanisms will be discussed and used in evaluating the experimental results.

2 Experimental

2.1 Materials

Epoxy resin Epon 828, a bisphenol A diglycidyl ether-based resin made by the Shell Chemical Co., was chosen as the control epoxy resin. Its average molecular weight is about 380 g/mol. Bis(4-aminocyclohexyl)methane (PACM-20) was the curing agent. Copolymers with varying weight percentages of dimethyl, methyl trifluoropropyl (TFP), and diphenyl (DP) siloxane with a controlled molecular weight of approximately 2200 g/mol were prepared from dimethyl siloxane tetramer, methyl trifluoropropyl cyclic trimer, and diphenyl tetramer via equilibration reactions with base catalysts, in particular tetramethylammonium siloxanolate and potassium siloxanolate. All siloxanes in this study had 2-aminoethylpiperazine (AEP) end groups. The oligomers were structurally characterized using FTIR and NMR spectroscopy. Number average molecular weights were determined by end group titration. Further details are given in Ref. [17].

The actual weight percentage of a specific type of siloxane unit in an oligomer can be calculated relative to the entire oligomer or the siloxane units alone. The number found by the first method, which includes the AEP end groups, will be smaller than that found by the second method. For example, in this work, we deal with two oligomers of essentially 100% TFP siloxane. If the end groups are included in calculations of percent TFP siloxane, the number assigned to the oligomer of the lower molecular weight is 70% while that for the higher molecular weight oligomer is 85%. Thus these end groups comprise an appreciable fraction of the oligomer molecular weight.

Siloxane-modified networks were prepared for testing via two steps. A linear precursor was generated by reacting the epoxy resin with the siloxane oligomer for one hour under vacuum at 65 °C. PACM-20 was then added, and the mixture was stirred for five minutes under vacuum at 50 °C. Previous studies indicated [15] that reaction between the AEP-terminated siloxane oligomers and the curing agent is not possible, as one would expect.

The epoxy/siloxane/PACM-20 mixture was poured into a hot (120 °C) RTV-silicone mold of the precise shapes to be used for solid-state testing. The mixture was cured at 160 °C for 2.5 hours. The curing time and temperature chosen were considered to provide enough mobility for network formation. This conclusion was partially based on earlier studies which found a glass transition temperature of 150 °C for Epon 828/PACM-20 [35].

The B. F. Goodrich Company made available ATBN and CTBN rubbers of both low (10%) and high (17–18%) acrylonitrile (AN) content. Nominal molecular weights of these oligomers were about 3700. The ATBN oligomers are manufactured from CTBN oligomers of the same AN content via capping with AEP. The ATBN materials should have molecular weights just slightly higher than the corresponding CTBN oligomers, but, due to excess AEP remaining in the ATBN after capping, the *titrated* molecular weights of the ATBN oligomers are considerably lower than their reported molecular weights. It is the titrated molecular weight that has been used to determine the appropriate amount of curing agent for complete network formation.

All networks, regardless of the modifier, were prepared identically. Hence, all modified resins were subjected to the same guidelines concerning the ratio of rubber to resin to curing agent and were cured according to the same schedule.

Glass transition temperatures were determined using a Perkin-Elmer DSC-II or DSC-IV at a heating rate of 10 °C/min. Several scans were run at 20 °C/min on a DSC-IV to gain information on the breadth of the glass transition region.

2.2 Sample Designation

Each oligomer or modified resin can be described by two or three numbers, respectively. The box below illustrates the employed shorthand nomenclature. For example, a sample containing 10% by weight of a 40% TFP siloxane rubber of molecular weight 2070 g/mol would be designated 10-2070-40F. Differences between specific types of siloxane copolymers are given by changes in the third number of the sample code while differences between the end groups of the butadiene oligomers are given in the first number of the code as demonstrated by the examples below.

Nomenclature
Wt. Percent — Modifier Mol. Wt. — Wt. % Comonomer
5, 10, 15   — 1500–4000   — TFP, DP, AN

EXAMPLES

Pure PDMS: 5-2190-0
TFP Siloxane: 10-2070-40F
Diphenyl Siloxane: 5-2250-20D
CTBN: 10C-3880-17AN
ATBN: 15A-2560-10AN

2.3 Static Mechanical Properties

Tensile moduli were measured from standard dog-bone samples (2.0 mm thickness, 4.7 mm width, and 22.0 mm gauge length) in a Model 1122 Instron. Flexural modulus was determined using a testing apparatus which consists of two aluminium/steel pieces attached to the Instron which is fitted with a tensile load cell. This device effectively performs an inverted three-point bend; the two side bars remain stationary above the sample as the central bar below the sample moves upward. Flexural samples measured ca. $52.0 \times 1.7 \times 13.1$ mm and were tested using a 25.4 mm span (distance between the two side bars). Crosshead speed (CHS) for both flexural and tensile testing was 1.0 mm/min.

Fracture toughness was measured in a three-point bend (3PB) geometry for all materials and in a compact tension (CT) geometry for a select group of materials. Dimensions of each sample, the location of the crack and crack notch, and the orientation of the testing direction are given in Fig. 1. Into the indicated notch was placed a sharp onesided razor blade which was tapped lightly to make a short pre-crack. The CHS was 0.5 mm/min for both geometries. After fracture, the precracks were

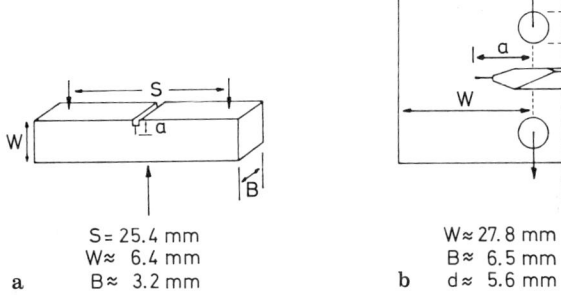

Fig. 1a and b. Fracture toughness test specimens: **a** three-point bend and **b** compact tension. Pre-crack length denoted by a. Movement of crosshead is vertical for both

enlarged with a magnifying glass, measured with vernier calipers, and calculated as the average of three values: side, center, side. The calculation of the fracture toughness will be described later in this chapter. All modulus and fracture toughness testing was done at ambient temperatures.

2.4 Dynamic Mechanical Properties

The variation of the damping factor (tan δ) with temperature was measured using a Polymer Laboratories Dynamic Mechanical Thermal Analyzer (DMTA). The measurements were performed on the siloxane-modified epoxies over a temperature range of $-150°$ to 200 °C at a heating rate of 5 °C per minute and a frequency of 1 Hz. The sample dimensions were the same as those used for flexural modulus test specimens.

2.5 Scanning Electron Microscopy

Fracture surfaces were examined in an ISI, Inc. Super III-A SEM. Cold snap samples were fractured after being submerged in liquid nitrogen for ten minutes. Crack faces of the 3PB and CT specimens were also studied with SEM.

2.6 Friction and Wear

Two wear tests were selected. In one test, fatigue was expected to be the predominant wear mechanism; in the second test, abrasive wear and adhesive transfer were expected to predominate. In the fatigue test, a 3.125 mm diameter stell ball bearing was loaded against the free air surface of a cast epoxy disk which was rotated to produce a sliding velocity of 0.63 m/s. The normal loads, 2, 5, and 10 N, were selected so that the calculated Hertzian elastic stresses in the sample were below its fracture strength. Abrasive wear was minimized because the ball surface was very smooth and the stresses in the epoxy were in the elastic range. Wear did not occur as soon as the test was started because the level of the stresses in the sample surface was low. After an initiation period, however, a wear groove formed in the surface of the sample which enlarged gradually throughout the test. After every 2000 revolutions of the sample, the test was suspended and the profile of the wear groove was measured with a surface profile meter. The slope of a linear regression of the wear track cross-sectional area with the number of disk rotations was taken as the wear rate expressed in units of area/cycle. Friction was measured continuously during the test. Epoxies modified with the dimethyl siloxane oligomer, dimethyl-methyl trifluoropropyl siloxane co-oligomer [36], and the ATBN and CTBN rubbers were used in the test.

In the second test, 2.2 mm diameter cylindrical pins were machined from the modified epoxy disks so that the test end of the cylinder was the original free air surface of the cast samples. The pins were loaded against a randomly abraded steel disk or against a smooth glass disk. The loads and speeds used were the same as those used in the fatigue tests. Friction and the decrease in the length of the pin as it wore were measured continuously; the wear rate was reported as the decrease in the length per unit sliding distance, m/m. The change in the appearance of the pin end while sliding on the glass disk was observed with an optical microscope. Tests were performed on a limited number of siloxane-modified epoxies [37].

3 Results and Discussion

3.1 Characteristics of the Siloxane-Modified Epoxy Networks

The synthesis of the siloxane-modified networks involved two steps as illustrated in Fig. 2. In the first step, the piperazine-terminated polysiloxane oligomer (at a concentration of 5–15 wt.-% of the total network) was allowed to react with an excess of Epon 828. Reaction occurred between the secondary amine piperazine end groups of the polysiloxanes (functionality, $f = 2$) and the epoxide rings of Epon 828. The reaction mixture after the completion of the first step contained excess unreacted Epon 828 and linear precursors resulting from the capping of the siloxane oligomer with Epon 828. In the second step, the final network was formed by reaction between the primary amine end groups of PACM-20 (functionality, $f = 4$) and the epoxide end groups on both Epon 828 and siloxane oligomers capped with Epon 828. It should be emphasized that the siloxane rubber modifier is chemically bonded to the epoxy matrix.

The control Epon 828/PACM-20 network was formed from a 2:1 molar ratio of Epon 828 to PACM-20. For the siloxane-modified networks, the molar ratio of the reactants was controlled such that the number of moles of epoxide groups in Epon 828 was equal to the combined mole number of N—H groups from the primary amines of PACM-20 and the piperazine amines of the polysiloxane modifiers.

The completeness of the crosslinking reaction can be shown by FTIR spectra such as that given in Fig. 3. Spectra are shown for a mixture of epoxy resin and curing agent immediately after being mixed and after being cured at 160 °C for 2.5 hours. The epoxide ring absorbs at 915 wavenumbers (cm^{-1}) and appears to overlap with a weakly absorbing band at approximately the same position. This latter band does not show any significant change after cure. It can be seen that the network was completely cured as judged by the disappearance of the narrow epoxide peak and the appear-

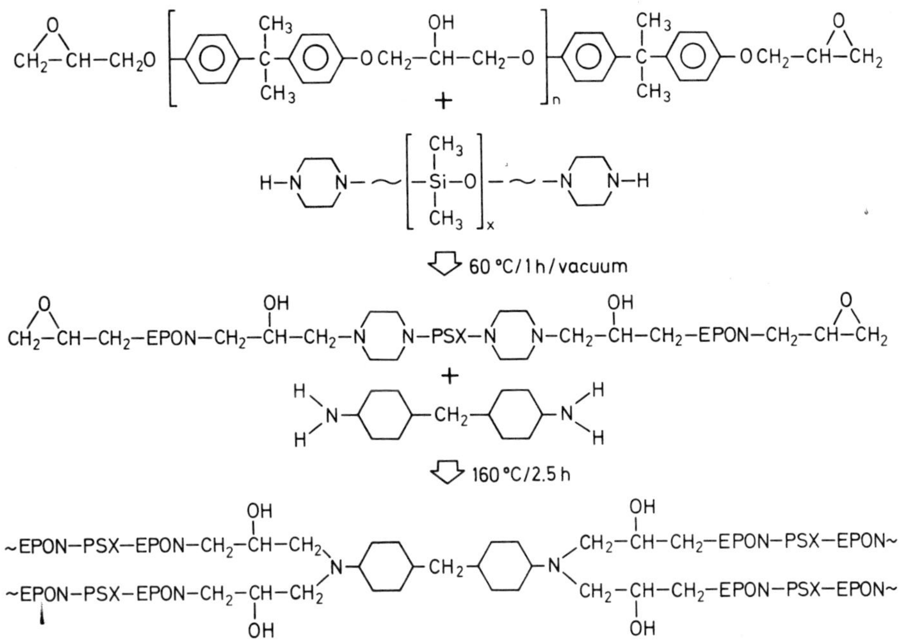

Fig. 2. Two-step synthesis of siloxane-modified epoxy network. First step, capping of siloxane oligomer with Epon 828. Second step, crosslinking of Epon 828 and capped siloxane with PACM-20

ance of a broad OH peak at about 3500 cm^{-1}. Longer cure times produced no further change in the FTIR spectra at 915 cm^{-1} and no increase in T_g beyond 150 °C.

In addition to the PDMS elastomer depicted in Fig. 2, we have produced for these investigations two series of siloxane copolymers based on dimethylsiloxane: 1) copolymers of methyl trifluoropropyl (TFP) siloxane and dimethyl siloxane and 2) copolymers of diphenyl (DP) siloxane and dimethyl siloxane. Figure 4 provides structures for the oligomers, all of which had aminoethylpiperazine (AEP) end groups. Details on synthesis and chemical characterization of these modifiers will be published [38].

It is believed that inclusion of the more polar TFP unit and the more resonant DP unit can allow one to control the compatibility between epoxy and elastomer. In the case of TFP siloxane, one of the two methyl units of dimethyl siloxane is replaced with a TFP group whose strong electronegativity creates a dipole across an otherwise non-polar chain. The epoxy resin is more strongly attracted to this new polar elastomer, and the more TFP incorporated, the greater the affinity between resin and rubber. Therefore, as cure proceeds, a siloxane of higher TFP content will precipitate out later in the crosslinking process. It is possible that some siloxane of high TFP content could remain dissolved in the epoxy matrix after cure is complete. Since the mobility of the network system diminishes continuously with cure time, at the point where incompatibility is reached, the elastomer segments of each chain (see Fig. 2) can coalesce with only so many other such segments. The net result is a particle size that is inversely proportional to TFP content. This argument can be applied to the effects of increasing DP content relative to dimethyl content in a siloxane modifier. In this

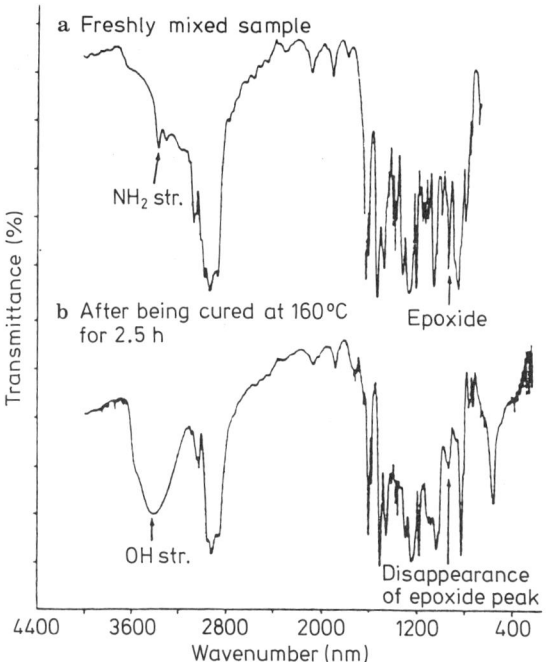

Fig. 3. FTIR spectra of Epon 828/PACM-20 before and after standard curing schedule. Taken on Nicolet MX-1 Spectrometer

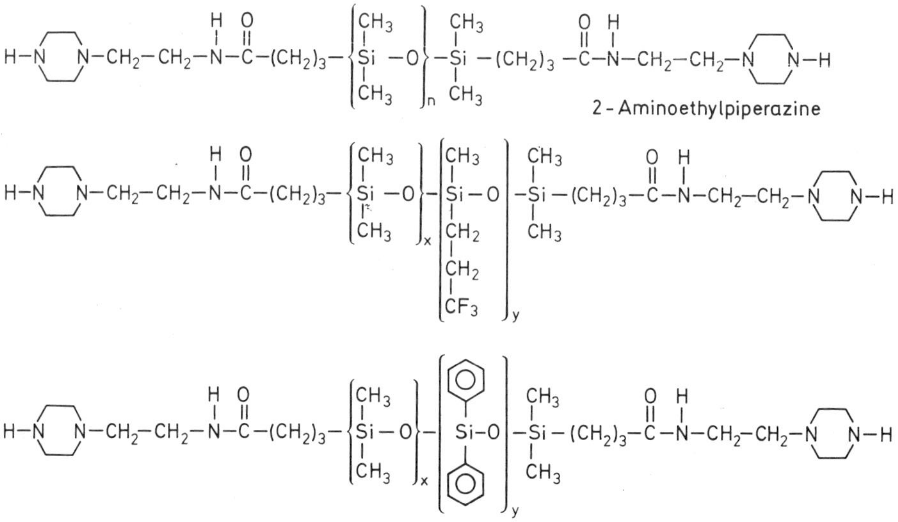

Fig. 4. Siloxane oligomers used for epoxy modification. From top, poly(dimethyl siloxane), poly(dimethyl-co-methyltrifluoropropyl siloxane), and poly(dimethyl-co-diphenyl siloxane). Aminoethylpiperazine is the functional end group

Table 1. Glass Transition Temperatures of Siloxane Oligomers

Oligomer[a]	Wt.-% Comonomer[b]	T_g (°C)
2190-0	0	−126
2330-20F	25	−116
2070-40F	50	−91
3130-85F	100	−58
1500-70F	100	−45
2250-20D	25	−88
2290-40D	50	−51

T_g values based on DSC scanning rate of 10 °C/min.
[a] Wt.-% comonomer based on total oligomeric weight;
[b] Wt.-% comonomer without end groups

second case, the diphenyl unit interacts with the resin through its resonant phenyl rings. The preceding explanation also explains the consequences of raising acrylonitrile content in ATBN and CTBN oligomers. For any of these three cases, it is imperative that the resin, curing agent, and chosen elastomer be well-mixed before this sequence of events can take place.

Table 1 lists the glass transition temperatures for the pertinent siloxane oligomers as a function of TFP and DP contents. The percent of each comonomer is recorded with reference to the siloxane units as well as the entire oligomer. One notes the difference that this creates between the two nominally "100%" TFP siloxanes of different molecular weight. Note also the higher T_g values for the DP series at equal weight percents, a factor which limits their ease of synthesis and may affect their mobility during cure.

In Table 2 are tabulated the T_g values for resins modified with the siloxane copolymers described in Table 1. Transition temperatures for samples modified with oligomers containing primarily dimethyl siloxane units give little indication of intimate mixing beween epoxy and rubber. Evidence for partial miscibility with increasing

Table–2. Glass Transition Temperatures of Epoxy Networks after Modification with Siloxane Oligomers

Oligomer	Wt.-% Rubber in Epoxy[a]		
	5	10	15
2190-0	158	153	152
2330-20F	157	147	151
2070-40F	150	136	150
3130-85F	150	150	150
1500-70F	151	145	140
2250-20D	149	150	148
2290-40D	159	154	140

[a] T_g of Epon 828/PACM-20 is 150 °C

TFP or diphenyl content is suggested by the relatively lower T_g's for some of the modified resins, in particular 10-2070-40F, 10- and 15-1500-70F, and 15-2290-40D.

The dynamic mechanical properties of the siloxane-modified epoxy networks were also investigated. The DMTA curves for the control epoxy network exhibit the two major relaxations observed in most epoxy polymers [39,40,41]. A high temperature or α transition at 150 °C corresponds to the major glass transition temperature of the network above which large chain motion takes place. The low temperature or β transition is a broad peak extending from −90° to 0 °C with a center near −40 °C. It has been attributed predominantly to the motion of the $CH_2-CH(OH)-CH_2-O$ (hydroxyether) group of the epoxy [39,40,42].

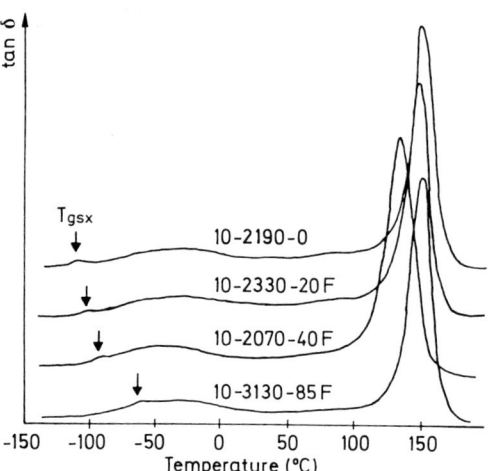

Fig. 5. Tan δ as a function of temperature for epoxy modified with 10 weight percent of dimethyl/TFP siloxane oligomers

Figure 5 shows the DMTA tan δ versus temperature curves of a series of siloxane-modified epoxy networks containing 10 wt.-% dimethyl-co-trifluoropropyl methyl siloxane oligomers of increasing TFP content. In addition to the epoxy α and β peaks, these curves show an additional small peak corresponding to the glass transition temperature of the siloxane phase. This small peak further supports the multiphase nature of the modified networks in which the epoxy resin and the elastomer are phase separated. It is interesting to note the gradual shift of the siloxane peaks (indicated by vertical arrows) to higher temperatures with increasing TFP content. The dependence of this transition peak on TFP content confirmed its assignment to the siloxane moiety.

The T_g values determined by DSC for the pure liquid siloxane oligomers were in good agreement with the values determined from DMTA of siloxane-modified epoxies. However, at 0 and 20% TFP content, the siloxane T_g from DMTA was about 16 °C *higher* than the T_g found by DSC. This suggests that at TFP contents above 20%, the siloxane separates from the epoxy as a "purer" phase. This point will be discussed further in the next section. Also reserved for later discussion is the depression of the major epoxy transition with an increase of the 2070-40F oligomer.

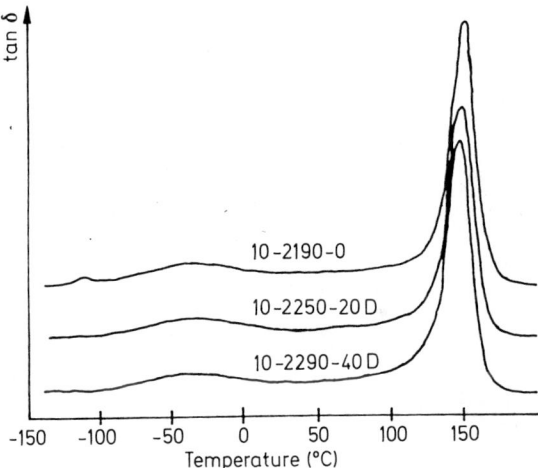

Fig. 6. Tan δ as a function of temperature for epoxy modified with 10 wt.-% of dimethyl/diphenyl siloxane oligomers

The damping curves of the epoxy networks modified with dimethyl-diphenyl siloxane oligomers are given in Fig. 6. The tan δ curves for these materials show two transitions which correspond to the glass transition and secondary relaxation of the epoxy network. The relatively strong β peak of the epoxy dominates that portion of the damping curve at which one would expect to see the transitions of the 2250-20D and 2290-40D oligomers (see Table 1). However, these transitions would be expected to be weak since these oligomers contain, on the average, only 2.2 or 4.6 diphenyl siloxane units per 17 or 12 dimethyl siloxane units, respectively.

3.2 Morphology and Solid-State Properties

Morphological investigations clearly illustrate the enhancement of resin-rubber compatibility through siloxane copolymerization. SEM photomicrographs of cold snap surfaces for the control and four TFP siloxane-modified resins are given in Fig. 7. Increasing TFP content decreases particle size from about 50 to less than 5 microns. The make-up of the particles also changes. At low TFP content, the relatively large domains have a nodular character. Transmission electron microscopy of CTBN-modified epoxies [25-27] indicates that such particles are actually mixtures of resin and rubber although that same conclusion cannot necessarily be made in this case. It is suspected that these particles result from incomplete mixing caused by gross incompatibility between the liquid elastomer and liquid resin prior to cure. At the highest TFP contents, particles are smaller and homogeneous in texture. As will be seen, domains less than 1 μm result from modification with 2250-20D and especially 2290-40D. Domains are roughly textured for both diphenyl siloxane modifiers, becoming irregularly shaped in 15-2290-40D.

A primary effect of increasing siloxane molecular weight is illustrated in Fig. 8. Small presumably homogeneous domains characterize the 15-1500-70F fracture surface (upper micrograph). When molecular weight for a pure TFP siloxane is doubled

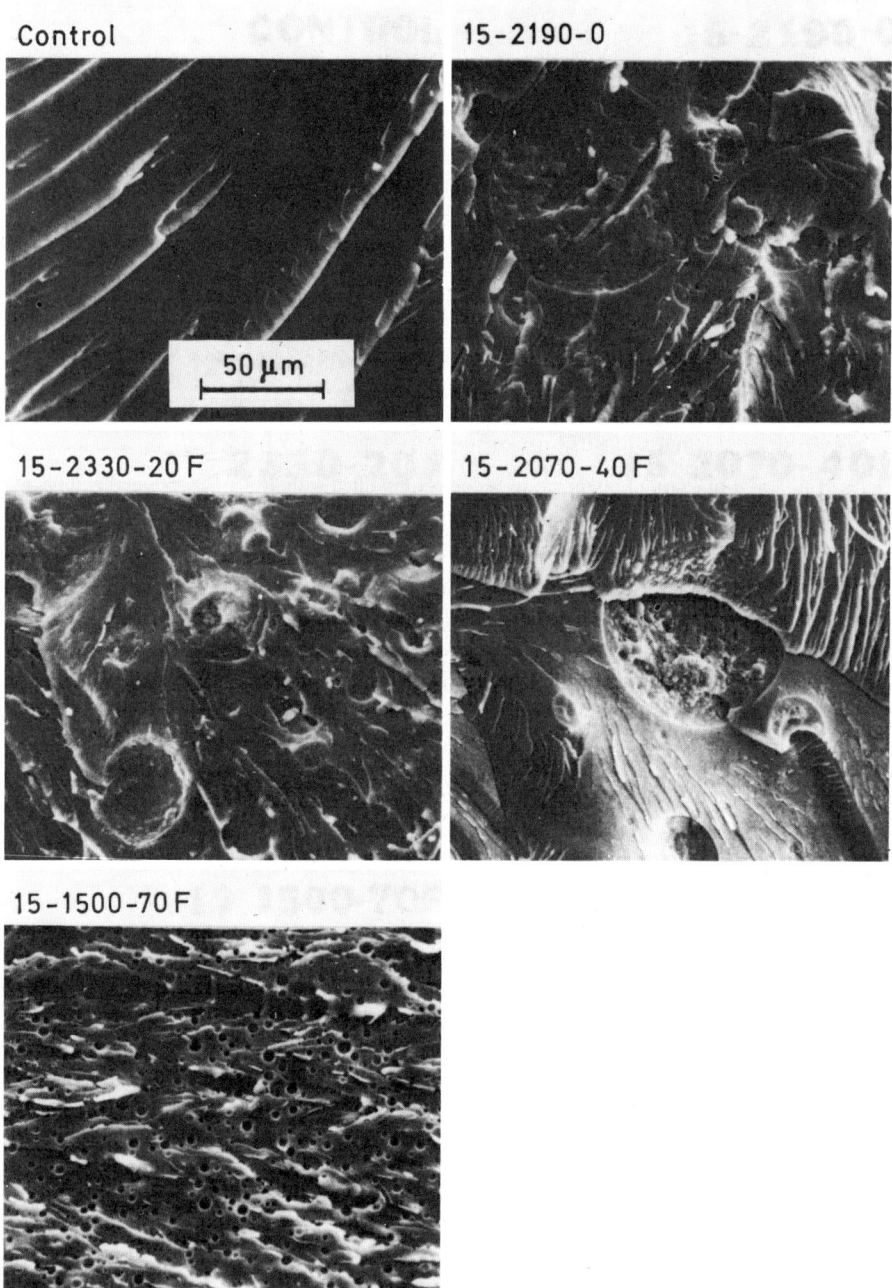

Fig. 7. Cold snap fracture surfaces for unmodified resin and four siloxane-modified resins as a function of TFP siloxane content in modifier. Original SEM magnification, 300×

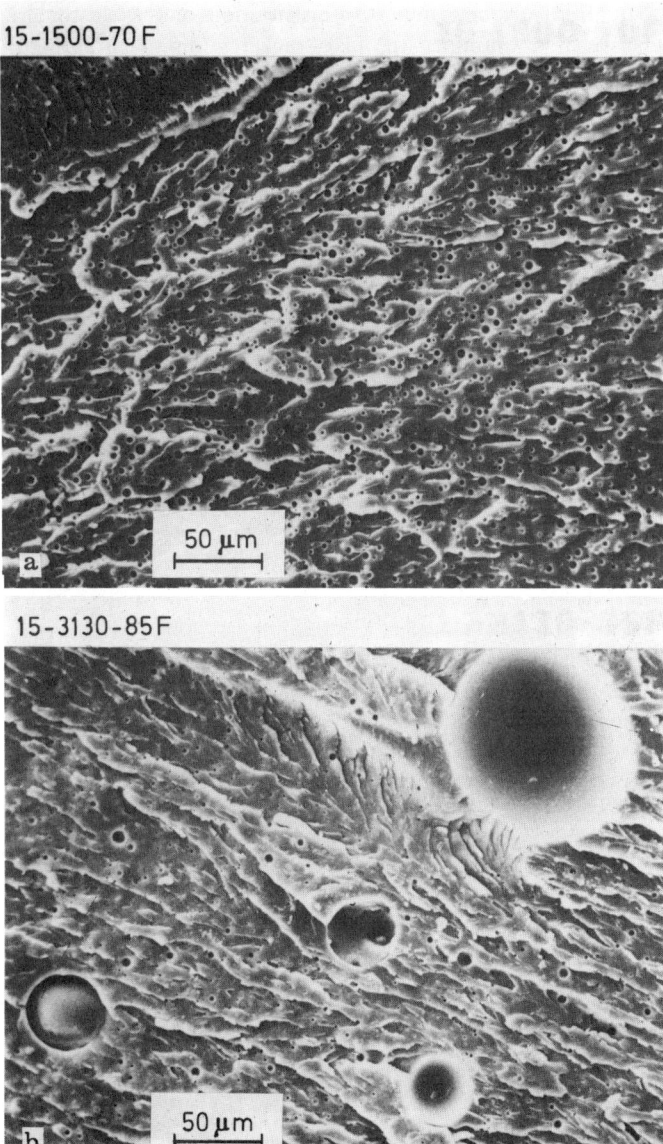

Fig. 8. Cold snap fracture surfaces showing effect of molecular weight of modifier on morphology. Modifiers in top and bottom micrographs are nominally 100% methyltrifluoropropyl siloxane. Original SEM magnification, 300 ×

(Refer to Table 1), the cold snap fracture surface shows a few large particles as well as small homogeneous uniformly textured domains.

Although it was demonstrated that the siloxane component of these materials migrates to the surface (top 50 Å) during cure [15], there is no evidence that this creates a concentration gradient of siloxane throughout the entire sample thickness. However, in an effort to at least qualitatively determine the distribution of the elastomeric phase in these materials, we have examined the fast crack regions of 10-2330-20F and 10-2190-0 fracture toughness samples. These smooth cleaved surfaces contain relatively few fracture artifacts and thus serve as reasonable cross-sections. They indicate that from one outside edge to the other, there exists an apparently even distribution of the elastomeric phase.

A toughened material, by definition, features improvements in fracture resistance without substantial loss of mechanical strength or modulus. Figures 9a and 9b

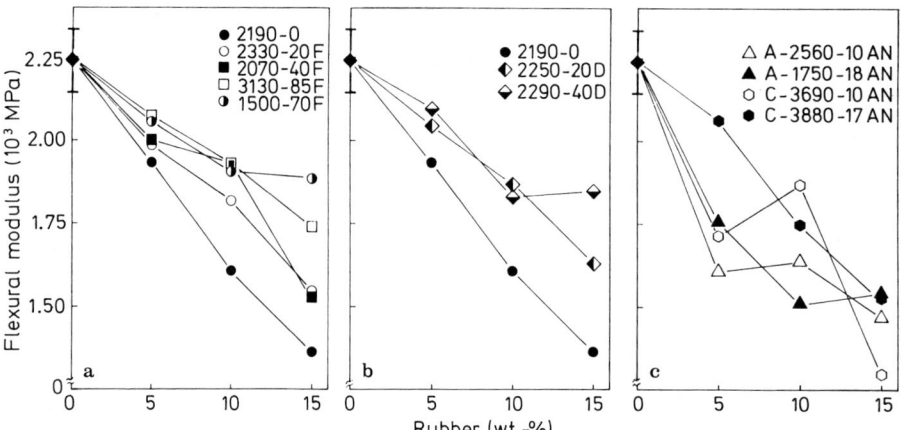

Fig. 9a–c. Flexural modulus as a function of rubber content and rubber composition: **a** TFP siloxanes, **b** DP siloxanes, **c** ATBN and CTBN. Error bars on data point for control are typical for all other points

illustrate that modification with the various siloxane oligomers only slightly influenced the flexural modulus of the base epoxy resin. As expected, the flexural modulus does decrease as rubber content increases. The decrease is less severe as either TFP or DP content increases. Although the T_g of the TFP and DP oligomers increases dramatically with TFP or DP content, it is not likely that the difference in T_g between the various rubbers bears heavily on the modulus since modulus testing is done at room temperature.

Consideration of the nature of the phase separation in siloxane modified epoxies permits a second explanation of these modulus trends. The morphologies observed contained elastomeric domains of a variety of sizes and textures [43] which changed with TFP content. At low TFP content especially, these particles are large and appear to be mixtures of epoxy and siloxane. If particularly large, about 50 µm in diameter, such domains were often observed on cold snap fracture surfaces to be weakly attached to the matrix such that gaps or voids existed between particle and matrix. Voids are well known to detract from modulus. Secondly, if the supposition that these large particles contain both elastomer and epoxy is correct, then the actual volume fraction of the elastomeric phase is larger than what would be calculated assuming the phase separation of "pure" siloxane phase without any epoxy inclusions. It is the volume fraction of the phase separated elastomeric domains which is believed to proportionately lower modulus [25]. In addition, if raising TFP content encourages miscibility between the modifier and the matrix, as has been suggested [17, 38], the effective volume fraction of phase-separated elastomer will be lowered. Studies with CTBN-modified epoxies [25, 26] have suggested that dissolved rubber is not as harmful to room temperature modulus as phase-separated rubber. A third situation is that where a small and more or less homogeneous domain adheres well to the matrix but where its size varies with TFP content. The strength of the interface may then partially control the modulus. For example, the siloxane-siloxane contacts within a homogeneous rubber particle consist primarily of van der Waal's forces. In contrast, the forces across the siloxane-epoxy interface also include inherently stronger covalent bonds and possible dipole-dipole interactions. With increasing TFP, particles become smaller and surface-to-volume ratio goes up, and the resin-rubber interface may contribute to and help to improve the modulus.

In summary of these points, it is seen that the isolation of particles from the epoxy matrix, the effective volume fraction of the elastomeric phase, and strength of the interface interact to control modulus. The morphology which a particular siloxane modifier promotes determines the contribution of any or all of these three factors to the modulus of the modified resin.

Figure 9c provides flexural modulus data for the CTBN- and ATBN-modified epoxies prepared for this study. Once again, modulus decreases with increasing rubber content, and increasing AN content seems to have an effect similar to increasing TFP or DP content in the siloxanes. If one compares the results of Figs. 9a and 9b with Fig. 9c, which are drawn on identical scales, one may observe that the TFP and DP siloxane-modified epoxies generally have higher moduli than the butadiene-modified resins. While subtle, this observation is in fact supported by Young modulus data, a portion of which is given in Table 3.

These differences in modulus may be at least partially explained by DSC data such as that in Fig. 10. It is seen that in general the glass transition regions of the ATBN- and CTBN-modified epoxies are broader and have a lower midpoint than those of the control and two siloxane-modified materials. This thermal data suggests that the butadiene oligomers are relatively more miscible with the epoxy and may act as plasticizers. As an additional point, it is likely that the higher molecular weight of the

Table 3. Selected Young's Moduli

Sample	Young's Modulus (10^3 MPa)
Control	1.11 ± 0.17
15-2190-0	0.92 ± 0.09
15-2070-40F	0.95 ± 0.07
15C-3690-10AN	0.88 ± 0.02
15C-3880-17AN	0.86 ± 0.04
15A-2560-10AN	0.68 ± 0.06
15A-1750-18AN	0.77 ± 0.05

Crosshead speed = 1.0 mm/min

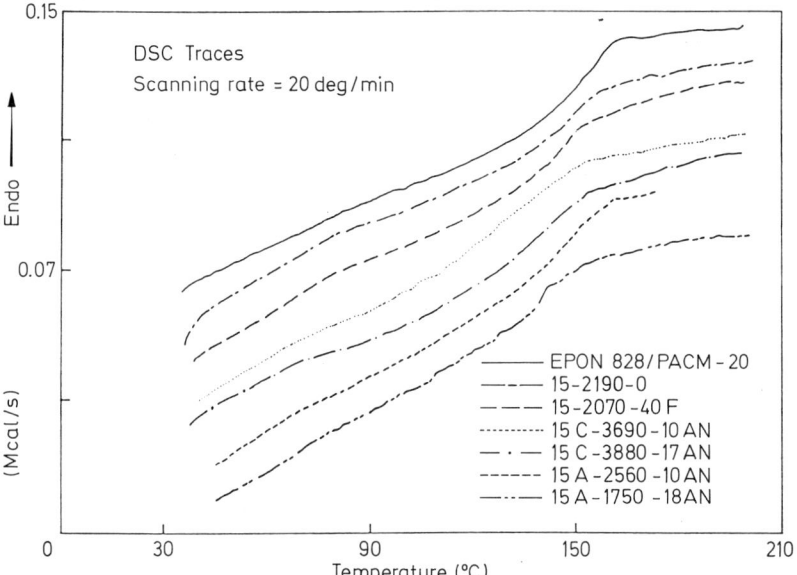

Fig. 10. Normalized DSC scans of resin control and six modified resins. Curves have been shifted vertically

CTBN oligomers leads to a higher molecular weight of chains between network junction points and consequently, a lower crosslink density.

In the only such study known to the authors, Sayre, Assink, and Lagasse [23] demonstrated (using energy dispersive x-ray analysis, NMR, and DSC) that the interface between the CTBN (18% AN) and epoxy phases in a diethanol amine-cured Epon 828 resin was not diffuse and that only a small percentage of CTBN was actually dissolved in the matrix. They also found that about one half of the CTBN precipitated in particles measurng less than 0.1 μm (1000 Angstroms) across. Although the curing agent and curing schedule for the resins studied by Sayre and co-workers were not the same as those used in the present study, their characterization indicates that fine mixing can occur between CTBN and Epon 828.

Fracture toughness for siloxane-modified as well as ATBN- and CTBN-modified resins was monitored through K_{IC}, plane-strain fracture toughness [44]. The K_{IC} values of at least five three-point bend (3PB) specimens of each material were calculated according to Eq. (1)

$$K_{IC} = \frac{6P}{BW^{1/2}} f(a/w) \tag{1}$$

where P is the critical load, B is the sample's width, W is its thickness, and a is the length of the pre-crack (See Fig. 1a). P is taken as the load at break. Letting R = a/w, the geometry factor Y = f(a/w) is as given in Eq. (2).

$$Y = 1.93R^{1/2} - 3.07R^{3/2} + 14.53R^{5/2} - 25.11R^{7/2} + 25.80R^{9/2} \tag{2}$$

A limited number of K_{IC} values were obtained with a compact tension geometry for which K_{IC} is calculated with Eq. (3)

$$K_{IC} = \frac{P}{BW^{1/2}} f(a/w) \qquad (3)$$

where symbol definitions are analogous to those for the 3PB test piece (See Fig. 1b). The geometry factor for the CT specimen is given in Eq. (4).

$$Y = 29.6R^{1/2} - 185.5R^{3/2} + 655.7R^{5/2} - 1017R^{7/2} + 638.9R^{9/2} \qquad (4)$$

The criteria of ASTM E399 (Ref. [44]) were followed as closely as possible. The only criterion which could not be consistently satisfied was that for a straight pre-crack, $\pm 5\%$.

In graphic presentation of K_{IC} results, the error bars given for the control are typical of all those data points which do not have their own error bars. In cases where error exceeded 10%, individual error bars are provided and labelled with the corresponding symbol. Such large deviations are thought to result from the violation of the homogeneity criterion of linear elastic fracture mechanics at 15% of certain oligomers. (See, for example, Fig. 7).

Fracture toughness results for the TFP siloxane-modified epoxies are given in Fig. 11a. With slight exception, modification with PDMS and 2330-20F gives virtually no improvement in K_{IC} and in fact lowers K_{IC} approximately linearly with weight

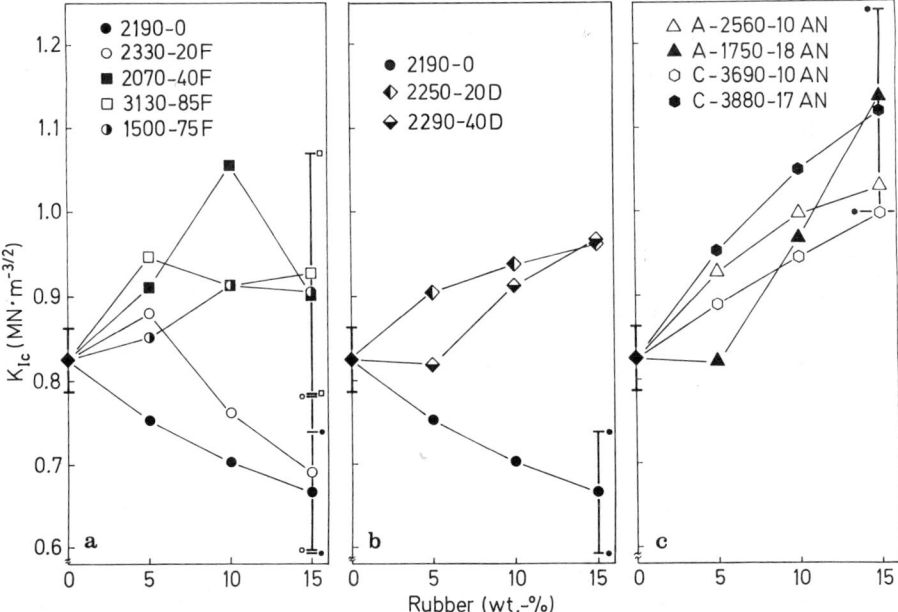

Fig. 11a–c. Fracture toughness as a function of rubber content and rubber composition: **a** TFP siloxanes, **b** DP siloxanes, **c** ATBN and CTBN. See text for error bar usage

percent of modifier! As TFP content climbs beyond 20%, K_{IC} increases considerably, reaching a high in the 10-2070-40F material. Although one might question this particular piece of data, scrutiny of the numerical data reveals percent errors of only 4.2, 3.5, and 7.1 for the series of epoxy modified with 2070-40F. Furthermore, the 10% sample showed a depressed resin T_g (see Table 2) of 136 °C followed by an increase to 150 °C at 15% rubber content, suggesting that the borderline between miscibility and immiscibility is located between 10 and 15% of 2070-40F. The DMTA curve for 10-2070-40F (see Fig. 5) also shows considerable depression of the epoxy glass transition temperature. It was seen earlier that the ATBN and CTBN modifiers also produce T_g depressions. Bucknall [28] has pointed out that a good toughening agent should be neither completely miscible nor immiscible with the material to be toughened.

Because the K_{IC} value is quantitatively derived from processes occurring in the pre-crack front, these were examined by SEM. Figure 12 shows two of the outstanding morphological features observed. The pre-crack front of a 10-2330-20F sample, which gave a K_{IC} value below that of the control (refer to Fig. 11a), is shown in Fig. 12a. One notes particularly the large heterogeneous particles of (presumably) both epoxy and elastomer. Rather than absorbing energy during loading, these particles rapidly tear, generally at an acute angle to the direction of crack propagation. The tearing observed here is not entirely unexpected in light of the fact that many siloxane elastomers have relatively poor tear resistance at room temperature [13,45], which may be related to the large interval between room temperature and T_g [46]. Kunz et al. [10,30] have postulated a toughening mechanism involving particle stretching (energy

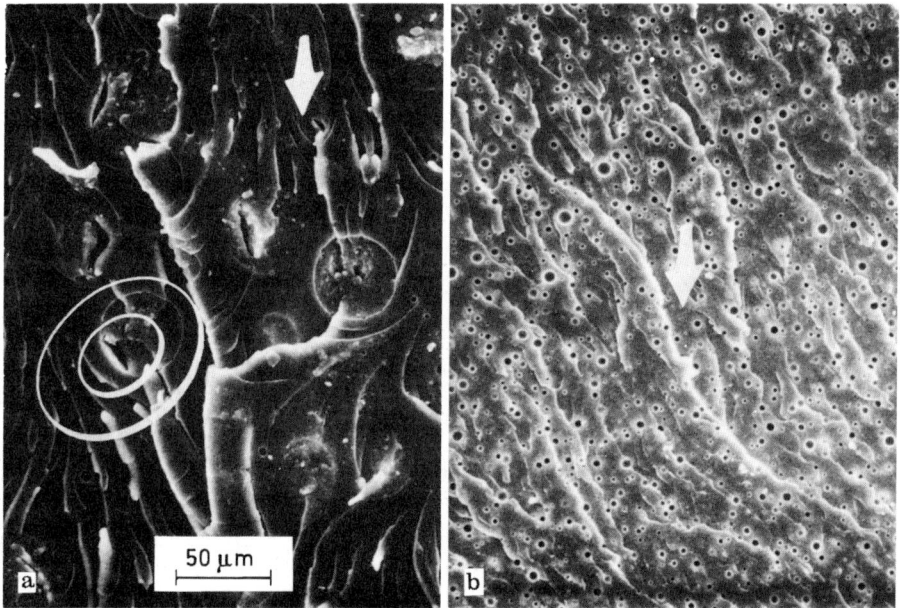

Fig. 12a and b. Pre-crack fronts of fracture toughness specimens: **a** 10-2330-20F and **b** 15-1500-70F. Ellipses highlight torn rubber domain. Arrows indicate direction of crack propagation. Original SEM magnification, 300 ×

absorption) and failure by tearing. However, the extent of tearing in this case and the associated K_{IC} values suggest that the tearing is not a significant part of an energy dissipative process. Interestingly, each tear is located either near the interface of a large heterogeneous particle or within such a particle which may itself be a maze of resin-rubber interfaces. These tears might therefore be regarded as primarily interfacial failures.

The second noteworthy morphological feature is presented in Fig. 12b. This micrograph depicts the K_{IC} pre-crack front of 15-1500-70F, which had a K_{IC} value significantly above that of the control, as shown in Fig. 11a. The holes may be examples of the dilatation effect observed in CTBN-modified epoxies [19, 22] in which rubber particles dilate in mutually perpendicular directions under the application of a triaxial stress and then collapse into spherical cavities following fracture. Dilatation requires a mismatch in coefficients of thermal expansion of resin and rubber [11]. This effect will therefore be most striking when the elastomeric phase is homogeneous, as is apparently the case here.

K_{IC} data for the DP siloxane-modified epoxies is provided in Fig. 11b. Even at

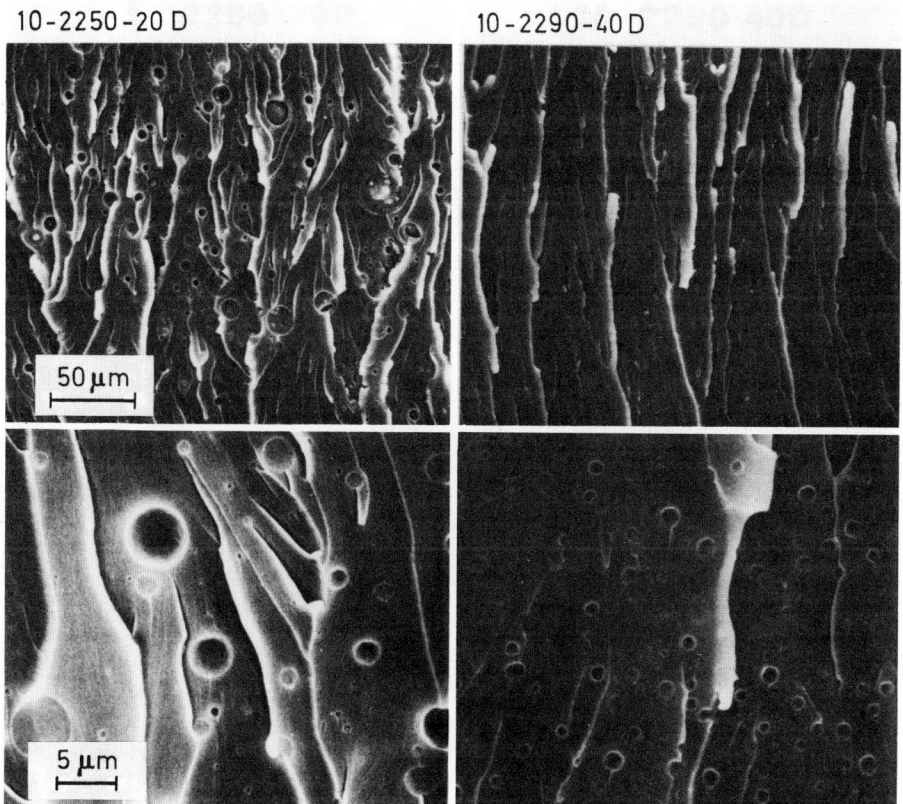

Fig. 13. Pre-crack fronts of fracture toughness specimens for resin modified with dimethyl/diphenyl siloxane oligomers. Original SEM magnifications, 300 × and 3000 ×

relatively low DP percentages, there is marked improvement in K_{Ic} over the control. As the micrographs in Fig. 13 show, the resins modified with these two oligomers exhibit smaller particle sizes than the TFP siloxane materials, however, the cavitation of the particles under fracture conditions is not as severe as in the TFP case. On the whole, the DP siloxane modifiers perform as well as or better than the TFP siloxane series, and studies of these modifiers are continuing.

One can now compare the K_{IC} improvement obtained with siloxane modification with that obtained using ATBN and CTBN modifiers, shown in Fig. 11c. (The specific results of Fig. 11c will be discussed later.) Looking collectively at the data of Fig. 11, it is seen that the 2070-40F oligomer is most competitive with the CTBN and ATBN oligomers at both 5 and 10% rubber content. The apparent partial miscibility of the 2070-40F oligomer is probably an important factor in this improvement.

Our interest in PDMS as an epoxy modifier lies partly in its low T_g relative to the ATBN and CTBN modifiers. Up to this time, however, improvements in K_{IC} through copolymerization of dimethyl siloxane with TFP and DP siloxane require raising the T_g of the siloxane modifier above that of PDMS, as shown by Table 1. It is hoped that increased understanding and control of the synthesis and morphologies of siloxane-modified epoxies will make it possible to retain the low T_g of the modifier while raising the fracture toughness of the resin. The true value of this objective could eventually be shown by measurement of K_{IC} at temperatures below ambient.

The fracture toughness results presented thus far were derived from a fairly small 3PB specimen. In limited work with a proportionately larger 3PB sample (B ~ 6.4 mm), a similar K_{IC} value was found for the control. A second geometry, compact tension (CT), was utilized for alternate determinations of K_{IC}. The K_{IC} values found with the CT specimens agreed within error with those obtained with the 3PB geometry, as shown in Table 4.

Table 4. Comparison of K_{Ic} Results from 3PB and CT Test Specimens

Sample	K_{Ic} (MNm$^{-3/2}$)	
	3PB	CT
Epon 828/PACM-20	.83 ± .04	.84 ± .04
10-2330-20F	.76 ± .06	.86 ± .05
10-2070-40F	1.06 ± .04	.95 ± .06

Study of the CT fracture surfaces by SEM reinforced earlier observations of the predominant features of fracture. The fracture surface of the unmodified epoxy was essentially featureless as before; a thin pre-crack front extended to a very smooth fast crack region. On the surface of the 10-2330-20F specimen observed there were found nodular 20–40 μm particles. In the pre-crack front, such particles exhibited irregular tears when crossed with a microcrack but otherwise remained undisturbed and encircled by an extremely fine border. In the fast crack region, slightly larger particles appeared to be more severely torn. Those small particles which did exist

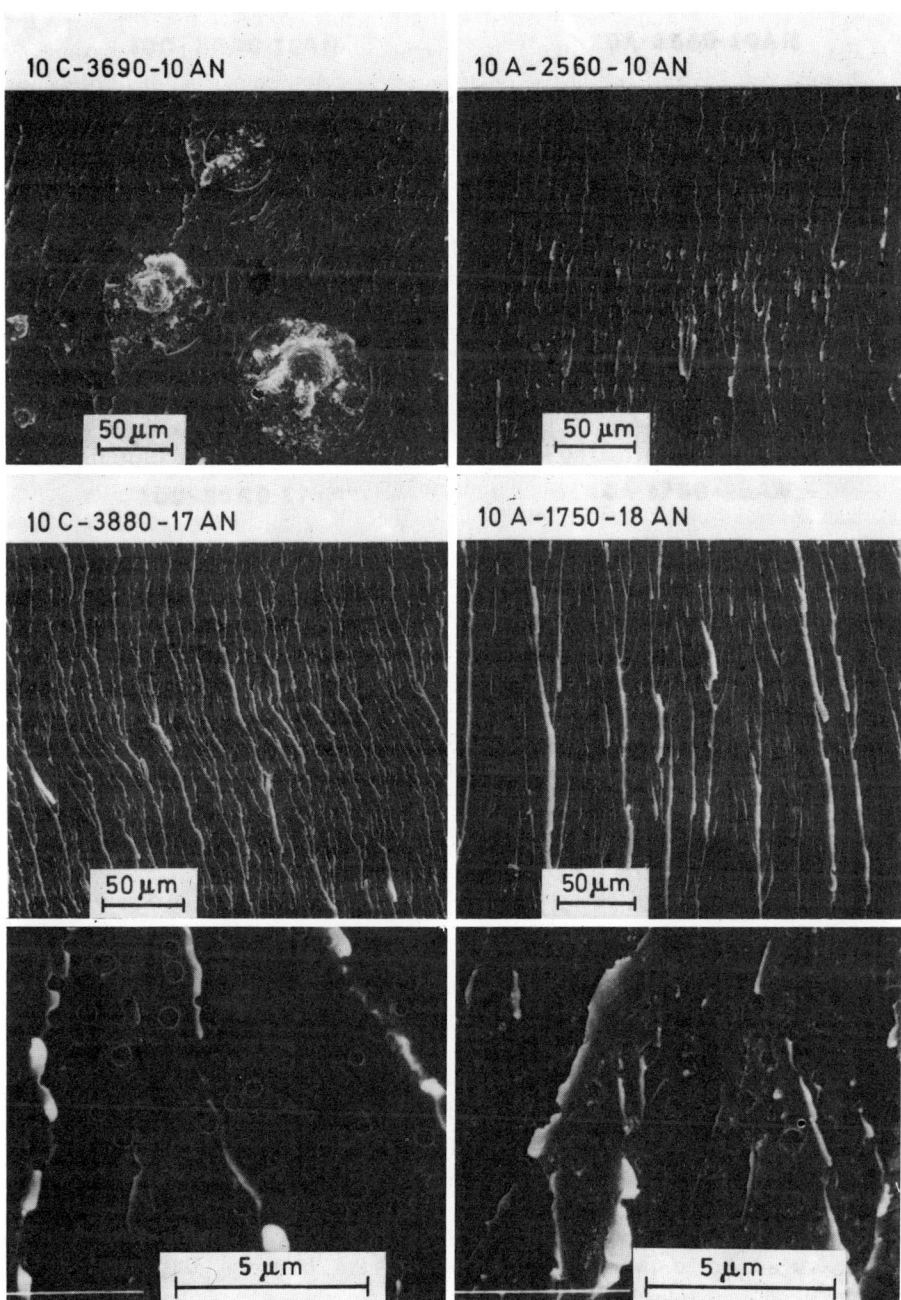

Fig. 14. Pre-crack fronts of resin modified with 10% of ATBN and CTBN elastomers. At top, low magnification micrographs of the four indicated materials. At bottom, high magnification micrographs of small particles in 10C-3880-17AN and 10A-1750-18AN. Original SEM magnifications, 300 × and 10000 ×

appeared to be essentially uninvolved in the fracture processes. In the pre-crack front of a 10-2070-40F CT sample, the average domain sizes were approximately 10–20 µm and 1–3 µm, the larger ones generally being somewhat heterogeneous in texture, the smaller ones presumably homogeneous. The larger domains in 10-2070-40F showed good adhesion to the epoxy matrix. The fast crack region showed similar features but finer microcracks. In general, the fracture surface of 10-2070-40F, which showed a K_{IC} value above those of both the control and 10-2330-20F, was rougher than 10-2330-20F, suggesting that a larger fraction of 10-2070-40F participates in the fracture, thereby allowing greater energy dissipation.

Looking strictly at the butadiene-based oligomeric modifiers, one sees from Fig. 11c that the CTBN-modified resins have higher K_{IC} values than the ATBN-modified resins at high AN content. At low AN content, the situation is exactly the opposite. Close examination of the ATBN and CTBN domains in K_{IC} fracture surfaces indicated that this reversal may have been caused by differences in particle morphology. Figure 14 points out these differences with four low magnification micrographs of resins modified with 10% of each of the four butadiene oligomers. Three of the samples show exclusively small-particle morphology while the 10C-3690-10AN surface contains large (25–50 µm) resin-rubber particles which coexist with small particles. The two lower micrographs in Fig. 14 illustrate the distinct difference between the small ATBN and CTBN particles. For the CTBN's of either AN content, the small particles are homogeneously textured and have sharp edges. The ATBN particles, though small for both high and low AN content, are somewhat nodular and may simply be small resin-rubber particles. These variations in morphology probably follow from the excess AEP in the ATBN, the different reactivities of the carboxyl and amine end groups, or a combination of these two factors. It is interesting to note that the small particles of either elastomer are roughly the same size. It is clear that when sizes are equivalent, more homogeneous domains promote greater fracture toughness.

3.3 Friction and Wear Properties

There are four generally recognized wear mechanisms: abrasive, fatigue, adhesive, and chemical. In abrasive wear, a harder material penetrates a softer one, and when sliding occurs, the hard material plows or chips wear particles from the soft one. The abrasive wear of polymers by harder materials has been found to be positively correlated with the inverse of the energy of the polymer to rupture [47]. If the stresses at contact points are lower than the fracture strength of the polymer, wear particles are not immediately generated when sliding commences. If the polymer surface is repeatedly stressed during sliding, fatigue cracks are eventually generated. These cracks propagate until wear particles are formed. Since fatigue wear is caused by repeated surface stresses, factors which reduce the stresses will decrease the magnitude of the fatigue. Hence, fatigue wear is reduced as the elastic modulus, normal load, and friction forces are reduced [48,49].

Adhesive wear is characterized by transfer of material from one material to another during sliding as a result of the adhesive forces at the contact points. The transferred material accumulates until the surface forces cause wear particles to form [50]. Because the rate at which polymeric material transfers depends on the bonding at the contact point — Coulombic and van der Waals — adhesive wear is usually small compared to abrasive wear. Adhesive wear can be isolated if the polymer is slid against a very smooth surface on which abrasive wear is eliminated [51]. Chemical wear occurs when the deformation losses cause temperatures to rise enough to cause degradation or softening of the polymer.

This brief review of friction and wear mechanisms indicates that the changes in fracture toughness, modulus, and surface energy induced by siloxane modification may cause changes in the friction and wear properties of siloxane-modified epoxy resins. Therefore, tests were performed to measure their friction and wear properties. In addition, some tests were run on the CTBN- and ATBN-modified epoxies for comparison purposes. A test designed to promote fatigue wear was run on all materials. A select group of materials was also subjected to a test during which adhesive and abrasive wear were expected to occur. The results of those tests will now be presented and discussed.

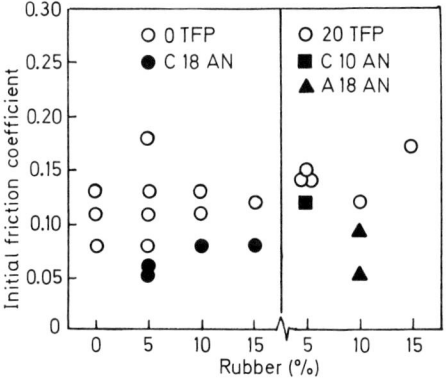

Fig. 15. Initial friction at 10 N load for the siloxane-, CTBN-, and ATBN-modified epoxies

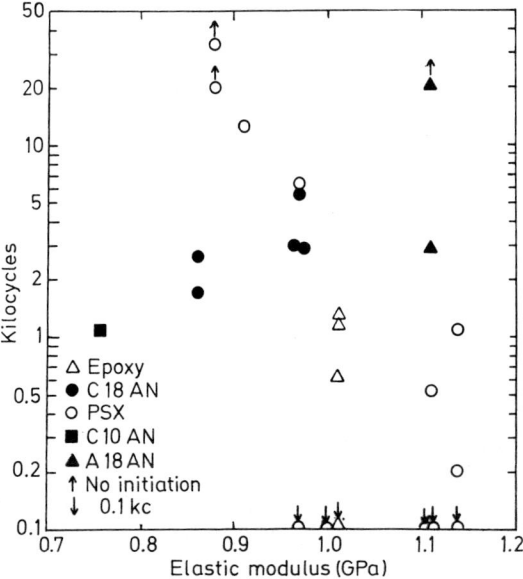

Fig. 16. Cycles to initiate wear at 10 N load as a function of elastic modulus

In the fatigue tests, the friction measured in the initial period of sliding during which no wear occurred was not affected significantly by the addition of siloxane or ATBN and CTBN modifiers to the epoxy as shown in Fig. 15 for 10N load. At lower loads, the friction coefficients are higher but still show no significant change with the additions of the modifiers.

The number of cycles of disk rotation required to initiate the wear track correlated positively with the weight percent of the siloxane modifier in the epoxy. However, the initiation times for the ATBN- and CTBN-modified epoxies showed no significant correlation with the percentage of the incorporated modifier. The initiation of the wear track is assumed to result from the fatigue of the epoxy; hence initiation time is related to the surface stresses. Because the surface stresses are inversely related to the elastic modulus as predicted by the Hertzian elastic contact theory [52], the initiation time data at 10N load were compared to the elastic moduli of the materials in Fig. 16. The initiation times for the siloxane-modified epoxies were negatively correlated with their elastic moduli while samples modified with ATBN and CTBN showed positive correlations with their moduli. At lower loads the initiation times for the siloxane-modified epoxies increased. The effect of load on the CTBN- and ATBN-modified epoxies was too erratic to show any significant trends.

Most tests were run for 14 kc of disk rotation. However, some materials did not initiate a wear track within the 14 kc, and the tests were extended, in some cases to over 30 kc. Even these extended tests were usually terminated before the wear track initiated. The modifier which produced the longest initiation times was the 20% TFP siloxane co-oligomer when present at 10 and 15 wt.-% in the epoxy.

Once the wear track was formed, the coefficient of friction increased two to three times due to two factors. One was the increase in contact area between the ball and the conforming wear track. A second was the development of wear debris which was

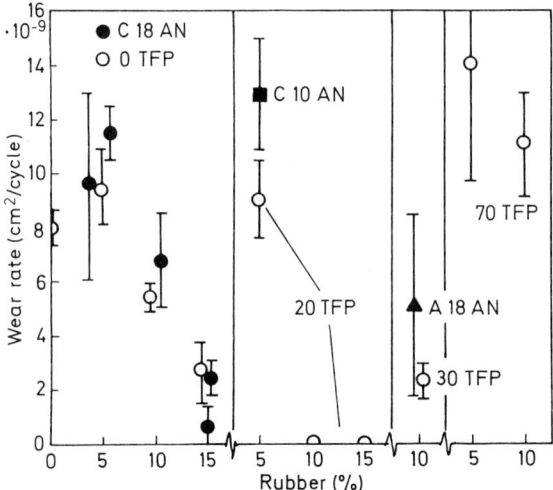

Fig. 17. Wear rates at 10N load for the siloxane-, CTBN-, and ATBN-modified epoxies

trapped between the ball and the track. The trapped debris either plowed or abraded the disk material or was pushed aside by the ball.

The wear rates of the modified epoxies are shown in Fig. 17 as a function of weight percentages of several siloxane and butadiene acrylonitrile modifiers. The two most significant results are the decrease in the wear rate which results from the increase in percentage of rubber and the minimum in the wear rate that occurs with the 20% TFP siloxane. The fatigue model for wear predicts that the wear will decrease as the elastic modulus decreases. Figure 18a indicates that this relationship holds for the siloxane and CTBN modifiers except for the 10% AN CTBN at 5 wt.-%. If the wear mechanism were predominantly abrasive, the wear rate should be negatively correlated with the energy to rupture of the materials. If K_{Ic} (as reported in the previous section) is used as the measure of fracture toughness, the data as shown in Fig. 18b indicate that the wear rates of the siloxane-modified epoxies correlated positively with the K_{Ic} values while those of the CTBN-modified epoxies correlated negatively with K_{Ic}.

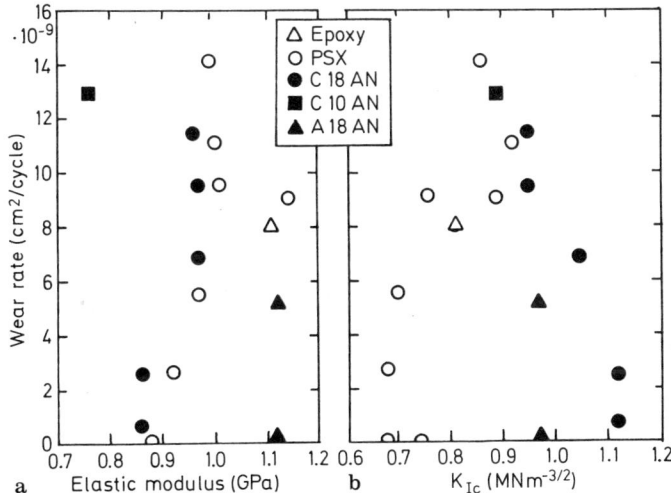

Fig. 18a and b. Wear rates at 10N load as a function of **a** elastic modulus and **b** fracture toughness

The tests in which the epoxy pins were rubbed on steel disks showed that the pins were initially worn by the abrasive action of the asperities on the steel surface. This initial wear correlated with the inverse of the K_{Ic} values. During the initial wear, the steel surface was smoothed by the transferred epoxy material. The steady state wear which followed the initial wear was lower in magnitude than the first stage of wear. The highest wear rate was obtained with 15 wt.-% of the dimethyl siloxane modifier.

The wear of the pins on the glass disk was observed with an optical microscope. The worn end of the pin was covered with material from the elastomeric domains. As the tests proceeded, the elastomer appeared to be rolled into thin cylinders oriented perpendicular to the sliding direction.

Most of the following discussion centers around the steel ball-on-epoxy disk experiments, for which a considerably larger body of data is available. In such an experiment, the friction during the initial period of sliding should be sensitive to changes in the surface energy if changes in surface energy reflect changes in the magnitude of tangential stresses at the contact points. The friction data indicated that there was no significant effect of the percentage of rubber regardless of the modifier. Thus it appears that the lower surface energy of the siloxane modifiers was negated by some factor. It could be that the inherent differences in surface energies were masked by species adsorbed during cleaning, handling, and storage. A second factor could be that epoxy is the predominant material in the effective surface for all materials tested; hence its adhesion to the steel ball dominates the total adhesion even at a 15 percent level of modifier.

The initiation of the wear track occurred as a result of failure of the epoxy material. Photomicrographs of the wear track formed at various times prior to wear track initiation showed the progression of events leading to wear particle formation. The first event was the appearance of transverse cracks in the center of the contact region. These cracks were a result of the tensile stresses in the direction of sliding. Repeated applications of these stresses on the cracks caused them to propagate, eventually creating loose wear particles. If rubber domains were large compared to the spacing between the domains then thin walls of epoxy material existed between the domains. These thin walls had lower modulus rubber domains on either side. When the steel ball slid over the thin walls, they failed as a result of bending stresses in the epoxy material.

Thus when examining the effect of the percentage of rubber of elastic modulus on the time required to initiate the wear track one must also consider the size and distribution of the domains on the surface. For the siloxane-modified epoxies, as the rubber content increased, so did the domain size. The domains in epoxy modified with the 20% TFP siloxane modifier covered the largest percentage of the surface. Above 20% TFP, the domain sizes decreased and were smallest at 70% TFP. The modulus values were lowest when the largest and most numerous domains were observed, i.e., for a given modifier, increasing weight percent lowered modulus and was accompanied by an increase in the size and number of elastomeric domains.

One sample containing the 20% TFP oligomer exhibited a wide range of domain sizes and distributions at its surface such that as the disk rotated, the ball encountered regions of different domain character. Whenever the wear track entered a region of closely spaced domains, the width of the track increased in response to the lower modulus in that region.

The CTBN- and ATBN-modified epoxies also had decreasing modulus values which were associated with increasing domain sizes. However, the time to initiate the wear track decreased as the elastic modulus decreased for these modifiers. A more detailed study of the wear track formation for these modifiers will be required to determine the reason for this relationship.

The modifier which gave the longest initiation time was the 20% TFP siloxane at levels of 10 and 15 wt.-%. This composition gave the size and distribution of domains which resulted in the optimum combination of reduction in modulus and spacing between domains. The reduction in modulus increased the area of contact and reduced the stresses on the surface. The spacing between the domains minimized the thin walls which would be subjected to bending failures. It must be emphasized that this optimization is dependent on the diameter of the ball chosen for the slider. A smaller or larger ball would result in different areas of contact and consequently different optimum domain sizes and spacings. Additional testing is required to determine the sensitivity of the initiation time to the interaction of the domain size and spacing with the ball size.

The initiation of the wear track occurred in the center of the contact area where the stresses were the highest. As soon as the epoxy wear particles started to form, the wear mechanism became a combination of abrasive wear by the particles and continued fatigue wear. The wear track widened until it covered the entire width of the contact area. During this period, the wear rate was highest; the rate then gradually decreased until it reached a steady state value. For 85 percent of the tests, the regression line of wear versus number of cycles of disk rotation had a correlation coefficient greater than 0.90, which gives an indication of the linearity of the data (i.e., the constancy of the wear rate).

The decrease in the wear rate with increasing rubber content and decreasing modulus of elasticity indicates that the predominant wear mechanism from this particular test is fatigue. If the dominant wear mechanism were abrasive wear, the wear would be expected to correlate with the negative of the energy to rupture. Figure 18b shows that the wear rates could be correlated with fracture toughness using a parabolic curve fit of data. However, there is no physical explanation for low wear at both low and high fracture toughness values. As noted in the discussion above, the assumption of mechanisms must also be based on the domain morphology. Among the data needed to confirm the wear mechanism assumptions are the wear rates of the compositions which did not initiate a wear track before the tests were terminated. Additional long term tests would be required to obtain this data.

The initial and steady state wear rates of the siloxane-modified epoxy pins on the steel disks correlated with the inverse of the K_{IC} values which agrees with previous abrasive wear tests [47]. The steady state wear rates on the smooth glass disks were comparable to those on the steel disks. Thus in both cases the wear mechanism is abrasive wear by the wear particles trapped in the interface between the pin end and the disk.

In summary, we can first say that there is no significant evidence that the low surface energy siloxane-modified epoxies reduce friction compared with the unmodified epoxy or the ATBN- and CTBN-modified epoxies. Based on the results of the steel ball-on-epoxy experiments, the most significant effect of the siloxane modifiers is the reduction of the elastic modulus associated with large closely spaced domains. The longer initiation times and lowest wear rates observed for the siloxane-modified epoxies were generally associated with a lower modulus. Epoxy modified with the CTBN of 18% AN content also showed lower wear rates with lower modulus but, in contrast with the siloxane-modified resins, had shorter initiation times with lower modulus.

4 Concluding Remarks

We have presented an evaluation of epoxy resins chemically modified with poly(dimethyl siloxane) as well as poly(dimethyl-co-diphenyl siloxane) and poly(dimethyl-co-methyltrifluoropropyl siloxane). The composition of the siloxane modifier, which controls the compatibility of the resin and the elastomer, has a profound effect on the morphology and the resulting modulus and fracture toughness of the modified resin. These three factors in turn strongly influence the friction and wear properties of the modified resins. Comparison of siloxane-modified epoxies with similarly prepared ATBN- and CTBN-modified epoxies using identical experimental techniques has been instructive. Additional studies of siloxane-modified epoxies are in progress.

5 References

1. Potter, W. G.: Epoxide Resins. New York: Springer 1970
2. Epoxy Resin Chemistry and Technology. May, C. A., Tanaka, Y. (eds.). New York: Marcel Dekker 1973
3. Epoxy Resin Chemistry. Bauer, R. S. (ed.). Am. Chem. Soc. Adv. Chem. Ser. *114* (1979)
4. McGarry, F. J.: Proc. Roy. Soc. Lond. A. *319*, 59 (1970)
5. McGarry, F. J., Sultan, J. N.: 24th Ann. Tech Conf., Reinf. Plast./Compos. Div., SPI, Sect. 11-B (1969)
6. Rowe, E. H., Siebert, A. R., Drake, R. S.: Mod. Plast. *47*, 110 (1970)
7. Rowe, E. H.: 24th Ann. Tech Conf., Reinf. Plast./Compos. Div., SPI, Sect. 11-A (1969)
8. Riew, C. K., Rowe, E. H., Siebert, A. R.: Am. Chem. Soc. Adv. Chem. Ser. *154*, 326 (1976)
9. Romanchik, W. A., Sohn, Geibel, J. F.: Am. Chem. Soc. Symp. Ser. *221*, 85 (1982)
10. Kunz, S. C., Beaumont, P. W. R.: J. Mater. Sci. *16*, 3141 (1981)
11. Kinloch, A. J., Shaw, S. J., Tod, D. A., Hunston, D. L.: Polymer *24*, 1341 (1983)
12. Okamoto, Y.: Polym. Eng. Sci. *23*, 222 (1983)
13. Warrick, E. L., Pierce, O. R., Polmanteer, K. E., Saam, J. C.: Rubber Chem. Tech. *52*, 437 (1979)
14. Cush, R. J., Winnan, H. W., in: Developments in Rubber Technology – 2. Whelan, A., Lee, K. S. (eds.). London: Applied Science 1981
15. Riffle, J. S., Yilgor, I., Banthia, A. K., Tran, C., Wilkes, G. L., McGrath, J. E.: Am. Chem. Soc. Symp. Ser. *221*, 21 (1982)
16. Yorkgitis, E. M., Tran, C., Eiss, N. S., Jr., Hu, T. Y., Yilgor, I., Wilkes, G. L., McGrath, J. E.: Am. Chem. Soc. Adv. Chem. Ser. *208*, 137 (1984)
17. Tran, C.: Ph. D. Thesis, Virginia Polytechnic Inst. & State Univ., Blacksburg (1984)
18. Bascom, W. D., Cottington, R. L.: J. Adhesion, 7, 333 (1976)
19. Bascom, W. D., Cottington, R. L., Jones, R. L., Peyser, P.: J. Appl. Polym. Sci., *19*, 2545 (1975)
20. Bascom, W. D., Hunston, D. L., in: Adhesion 6. Allen, K. W. (ed.). London: Applied Science 1982
21. Bascom, W. D., Hunston, D. L., Timmons, C. O.: Org. Coat. Plast. Chem. *38*, 179 (1978)
22. Bascom, W. D., Mostovoy, S.: Org. Coat. Plast. Chem. *38*, 152 (1978)
23. Sayre, J. A., Assink, R. A., Lagasse, R. R.: Polymer *22*, 87 (1981)
24. Kunz, S. C., Sayre, J. A., Assink, R. A.: Polymer, *23*, 1897 (1982)
25. Manzione, L. T., Gillham, J. K., McPherson, C. A.: J. Appl. Polym. Sci., *26*, 907 (1981)
26. Manzione, L. T., Gillham, J. K., McPherson, C. A.: J. Appl. Polym. Sci., *26*, 889 (1981)
27. Bucknall, C. B., Yoshii, T.: Br. Polym. J. *10*, 53 (1978)
28. Bucknall, C. B.: Toughened Plastics. London: Applied Science 1977
29. Yee, A. F., Pearson, R. A.: General Electric Co. NASA Contractor Report 3718. 1983
30. Kunz-Douglass, S. C., Beaumont, P. W. R., Ashby, M. F.: J. Mater. Sci. *15*, 1109 (1980)
31. Sultan, J. N., McGarry, F. J.: Polym. Eng. Sci. *13*, 29 (1973)
32. Kinloch, A. J., Shaw, S. J., Hunston, D. L.: Polymer *24*, 1355 (1983)
33. Meeks, S. C.: Polymer *15*, 675 (1974)
34. Johnson, K. L. in: Friction and Traction. Dowson, D. et al. (eds.). Guildford, England: Westbury House (1981)
35. Yilgor, I., Yilgor, E., Banthia, A. K., Wilkes, G. L., McGrath, J. E.: Polym. Bull. *4*, 323 (1981)
36. Hu, T. Y., Eiss, N. S., Jr., Yorkgitis, E. M., Wilkes, G. L., Tran, C., Yilgor, I., McGrath, J. E.: Polym. Mater. Sci. Eng. *49*, 508 (1983)
37. Eiss, N. S., Jr., Czichos, H., in: Wear of Materials 1985. New York: ASME 1985, in press

38. Tran, C., McGrath, J. E.: in preparation
39. Ochi, M., Ozazaki, M., Shimbo, M.: J. Polym. Sci.: Polym. Phys. Ed. *20*, 689 (1982)
40. Pogany, G. A.: Polymer *1*, 66 (1970)
41. Enns, J. B., Gillham, J. K.: J. Appl. Polym. Sci. *28*, 2567 (1983)
42. Chang, T. D., Carr, S. H., Brittain, J. O.: Polym. Eng. Sci. *22*, 1205 (1982)
43. Yorkgitis, E. M.: Ph. D. Thesis, Virginia Polytechnic Inst. & State Univ., Blacksburg (1985)
44. Knott, J. F.: Fundamentals of Fracture Mechanics. New York: Wiley 1973
45. Gent, A. N., Tobias, R. H.: Am. Chem. Soc. Symp. Ser. *193*, 367 (1982)
46. Gent, A. N., Pulford, C. T. R. in: Developments in Polymer Fracture — 1. Andrews, E. H. (ed.). London: Applied Science 1979
47. Lancaster, J. K. in: Polymer Science, a Materials Science Handbook. Jenkins, A. D. (ed.). North Holland Publishing 1972
48. Jain, V. K., Bahadur, S.: Wear *60*, 187 (1973)
49. Eiss, N. S., Jr.: Polym. Mater. Sci. Eng. *50*, 78 (1984)
50. Eiss, N. S., Jr., Bayraktaroglu, M.: ASLE Trans. *23*, 269 (1980)
51. Briscoe, B. J., Tabor, D. in: Polymer Surfaces. Clark, D. T., Feast, W. J. (eds.). New York: Wiley 1978
52. Timoshenko, S., Goodier, J. N.: Theory of Elasticity. 2nd edit. New York: McGraw-Hill 1951

Editor: K. Dušek
Received February 20, 1985

The Application of Differential Scanning Calorimetry (DSC) to the Study of Epoxy Resin Curing Reactions

John M. Barton
Materials and Structures Department, Royal Aircraft Establishment, Farnborough, Hampshire, UK

This review is on the use of differential scanning calorimetry as a method of monitoring and investigating the kinetics of epoxy resin curing reactions. Some instrumental and experimental aspects are discussed, including methods of analysing the kinetic data. A brief survey is made of epoxy resin curing reactions and results of DSC studies are reviewed. These results are concerned with the use of carboxylic acid anhydrides, primary and secondary amines, dicyanodiamide, and imidazoles as curing agents.

1 Introduction 112

2 The DSC Technique 112
   2.1 DSC Instruments 112
   2.2 Application of DSC to Epoxy Resin Cure 114
      2.2.1 Measurement of Glass Transition Temperature 114
      2.2.2 Exothermic Heat Flow Measurements 115
         2.2.2.1 Basic Assumptions 115
         2.2.2.2 The Baseline 116
      2.2.3 Treatment of Kinetic Data 117

3 Epoxy Resin Curing Reactions 120
   3.1 Carboxylic Acid Anhydride Cure 120
   3.2 Amine Addition and Base-catalysed Cure 123

4 Results of DSC Studies 126
   4.1 Anhydride Curing Agents 126
   4.2 Primary and Secondary Amines as Curing Agents 131
   4.3 Dicyanodiamide and Imidazoles as Curing Agents 144
      4.3.1 Dicyanodiamide 144
      4.3.2 Imidazoles 148

5 Conclusions 151

6 References 151

1 Introduction

Epoxy resins are widely used in diverse applications including surface coatings, printed circuit boards, the potting of electronic components, rigid foams, adhesives, and fibre-reinforced composites. In all of these applications a curing process is involved in which the monomeric or oligomeric polyfunctional epoxide is transformed into a crosslinked macromolecular structure. An understanding of these curing reactions, together with the availability of reliable methods of monitoring them, are important in order to obtain consistent products with the desired physical and mechanical properties. The control of the curing process is especially important in the case of structural materials which are required to fulfil a critical design specification.

The technique of differential scanning calorimetry was introduced in the form of commercial instruments during the early 1960s, and it has been found to provide a convenient and useful method of monitoring the course of exothermic reactions including those involved in the cure of epoxy resins. Its main advantages are the modest requirements in terms of sample size, of the order of milligrams, and that it can provide quantitative data on overall reaction kinetics, with relative speed and ease. In addition it can give measurements of thermal transitions such as the glass transition temperature which is associated with the degree of crosslinking or state of cure of a resin. The interpretation of the results does however require a critical approach which is not always sufficiently evident in published work.

In Chapter 2 the DSC technique is discussed in terms of instruments, experimental methods, and ways of analysing the kinetic data. Chapter 3 provides a brief summary of epoxy resin curing reactions. Results of studies on the application of DSC to the cure of epoxy resins are reviewed and discussed in Chapter 4. These results are concerned with the use of carboxylic acid anhydrides, primary and secondary amines, dicyanodiamide, and imidazoles as curing agents.

2 The DSC Technique

2.1 DSC Instruments

The Nomenclature Committee of the International Confederation for Thermal Analysis (ICTA) has defined DSC as a technique in which the difference in energy inputs into a substance and a reference material is measured as a function of temperature whilst the substance and reference material are subjected to a controlled temperature program. Two modes, power compensation DSC and heat flux DSC, can be distinguished depending on the method of measurement used[1]. The relationship of these techniques to classical differential thermal analysis (DTA) is discussed by MacKenzie[2].

The power compensation DSC instrument was first described by Watson et al.[3] and by O'Neill[4] and it was developed into a commercial instrument by the Perkin-Elmer Corporation. It utilises separate sample and reference holders of low thermal mass, with individual heaters and platinum thermometers, as shown schematically in Fig. 1. In addidion to controlling the average temperature the instrument employs a

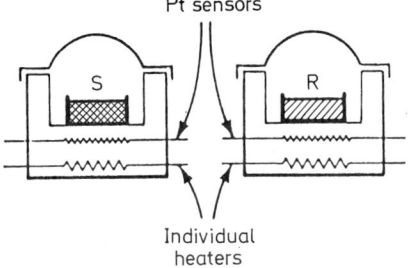

Fig. 1. Power Compensation DSC. Schematic Cross-Section of Perkin-Elmer DSC Cell. (Reproduces from Thermal Analysis Newsletter, Perkin-Elmer Corp., No. 9 (1970))

differential control loop which varies the power supplied to the individual heaters so as to minimise the temperature difference between the sample and reference thermometers. This power difference is proportional to the heat flow into the sample relative to the reference material.

Boersma [5] showed that quantitative calorimetric data could be obtained from a modified DTA instrument in which the sample and reference are in separate containers connected by a controlled thermal resistance, and with external thermocouples. In such an instrument the sample-reference temperature difference can be related to the heat flow, and this is the basis of heat flux DSC. The DuPont 910 DSC is based on a further development of this principle, and it is illustrated by Fig. 2.

This instrument utilises a silver block chamber with an external heater. The chamber contains a constantan disc with raised platforms for the sample and reference containers. The temperature difference between sample and reference is monitored by area thermocouples formed by the constantan disc and chromel wafers under the platforms. Amplification and electronic compensation of the differential temperature signal provides a linear calorimetric response over a wide temperature range. The theory of this instrument is discussed by Lee and Levy [6]. Other available examples of

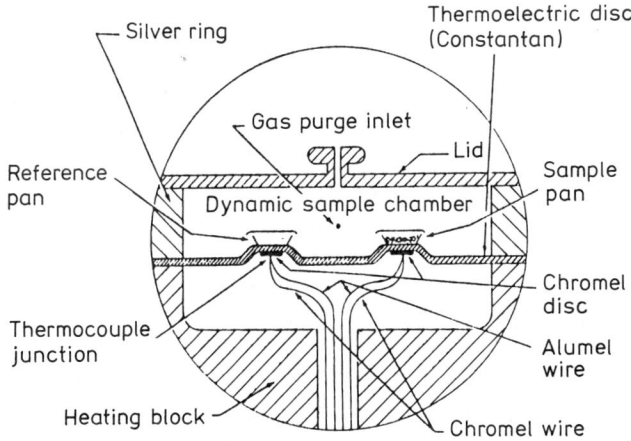

Fig. 2. Heat Flux DSC. Schematic Cross-Section of DuPont 910 DSC cell. (Reproduced from Product Bulletin, 910 DSC, DuPont Instruments, DuPont Co.)

heat flux DSC instruments include those manufactured by Linseis, Maple Instruments, Mettler, Netzsch, Setaram, and Stanton-Redcroft. A more comprehensive review is given by Wendlandt and Gallagher[7].

A comparison of the theoretical basis of the two types of DSC has been made by Mraw[8]. In both cases under conditions of dynamic equilibrium the output signal is proportional to the heating rate and to the difference in apparent heat capacity of the sample and reference materials. The output signal is calibrated in units of heat flow by means of a calorimetric standard such as alpha-alumina (enthalpy change) or a high purity crystalline metal (heat of fusion). Although there are basic differences in their mode of operation, both types of DSC can yield similar quantitative performance in the measurement of heat capacity and in monitoring heat flow during exothermic transitions or reactions [6, 9–10]. Small samples are necessary to prevent excessive temperature differentials [14], and good thermal contact is required between the sample, its container, and the platform.

The sample containers most commonly used are cylindrical pans pressed from pure aluminium foil. Alternative materials are used for very high temperatures or corrosive substances, and hermetically sealed pans to withstand several atmospheres pressure can be used for volatile materials. Some heat flux DSC instruments are available which are capable of operation at high pressures, by means of containment of the DSC cell within a pressure vessel.

2.2 Application of DSC to Epoxy Resin Cure

2.2.1 Measurement of Glass Transition Temperature

One of the major applications of DSC is the measurement of glass transition temperature (T_g) [11, 12]. In the absence of endothermic or exothermic reactions the DSC heat flow output is proportional to the sample heat capacity, and the T_g may be determined from the characteristic discontinuity in heat capacity. The T_g of a crosslinked polymer in general shows an increase with increasing degree of crosslinking, and thus provides a useful index of the degree of cure. The T_g is dependent on the chain flexibility and the free volume associated with the chemical structure as well as the overall crosslink density. An approximate empirical relationship between T_g and number average molecular weight between crosslinks (M_c) has been given by Nielsen [13, 14], from the correlation of data for a variety of network structures:

$$T_g - T_{g0} \simeq 39{,}000/M_c . \tag{2-1}$$

In this Eq. T_{g0} is the glass transition temperature of the uncrosslinked polymer having the same chemical composition as the crosslinked polymer.

Chompff [15] has developed a model for relating T_g to network structure which is based on free volume considerations. The network is treated as a terpolymer of branch points, chain segments, and chain ends, each providing a different free volume contribution. From the general model approximate solutions can be derived which agree fairly well with experimental data for some networks.

2.2.2 Exothermic Heat Flow Measurements

2.2.2.1 Basic Assumptions

A basic assumption in DSC kinetics is that heat flow relative to the instrumental baseline is proportional to the reaction rate. In the case of temperature scanning experiments the heat capacity of the sample contributes to the heat flow (endothermic), and this is compensated by the use of an appropriate baseline under the exo- or endothermic peak produced by the reaction. It is also assumed that the temperature gradient through the sample and the sample-reference temperature difference are small. Careful control of the sample size and shape, and the operating conditions are necessary in order to justify these assumptions.

A typical DSC scan for an exothermic reaction is shown schematically in Fig. 3.

For the present purposes a positive heat flow will be assigned to an exothermic event. The heating rate is fixed, so that there is a linear relationship between time and temperature. If there are m molecules reacting with a constant heat of reaction per molecule, then it is assumed that

$$dq/dt \propto -dm/dt . \qquad (2\text{-}2)$$

The constant of proportionality in Eq. (2-2) is Q_0/m_0 where Q_0 is the overall heat of reaction and m_0 it the initial number of reacting molecules, and

$$(dq/dt)/Q_0 = (-dm/dt)/m_0 . \qquad (2\text{-}3)$$

Integration of Eq. (2-3) implies that at any time, t, $Q_t/Q_0 = m_t/m_0$, where Q_t is the partial peak area:

$$Q_t = \int_0^t (dq/dt)\, dt . \qquad (2\text{-}4)$$

It is often convenient to work in terms of the fractional conversion, α, where $\alpha = Q_t/Q_0$ and

$$d\alpha/dt = (dq/dt)/Q_0 \qquad (2\text{-}5)$$

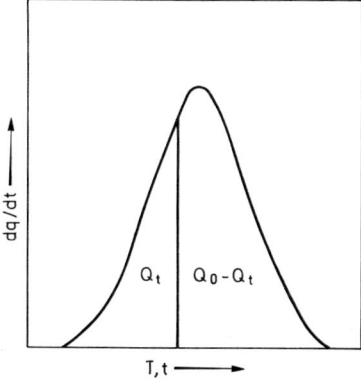

Fig. 3. Schematic DSC response for an exothermic reaction

2.2.2.2. The Baseline

The heat flow due to an exothermic reaction is given by the difference between the DSC output signal and its baseline response in the absence of reaction. An accurate baseline under the DSC thermogram is therefore necessary for quantitative work.

Two methods are commonly used to obtain isothermal data from DSC. The first method involves insertion of the sample into the DSC previously equilibrated at the required temperature. In the second method the sample is placed in the DSC cell at ambient temperature and the temperature is then increased at a controlled rapid rate to the required temperature. Small samples are used to ensure the sample temperature remain close to the required value. In both methods there is an initial off-balance signal and the output finally reaches a value corresponding to completion of the reaction. The baseline is usually taken as this final steady state signal, and horizontal negative extrapolation to intersect with the initial exotherm is taken as zero time for the reaction, as shown in Fig. 4.

At higher temperatures a significant amount of reaction may occur during the initial temperature ramp before approximate isothermal equilibrium has been attained. Some degree of correction for this is possible [16,17] by re-running the experiment on the reacted sample, under the same conditions, to obtain an estimate of the "true" baseline, as illustrated in Fig. 4.

For temperature scans the ideal case is when there is no significant change in heat capacity between reactants and products, and no changes in sample dimensions, so that the baseline is horizontal (Fig. 5a). Often there is an apparent change in heat capacity and then as a first approximation a sloping linear baseline is usually assumed, as in Fig. 5b. A better approximation may be obtained by assuming a baseline proportional to the fractional conversion, α, through the peak. This is illustrated schematically in Fig. 5c where the baseline deflection, $b(t)$, at a given time, t, is given by $b(t)/b_0 = Q(t)/Q_0 = \alpha(t)$, where b_0 and Q_0 are the overall changes in baseline shift and evolved heat. Initial values of b_0 and Q_0 are obtained from a linear sloping baseline, and a new baseline is calculated. The process is reiterated until there is a negligible change in peak area. Another problem is presented when a secondary exothermic process such as thermal degradation occurs towards the end of the reaction of interest.

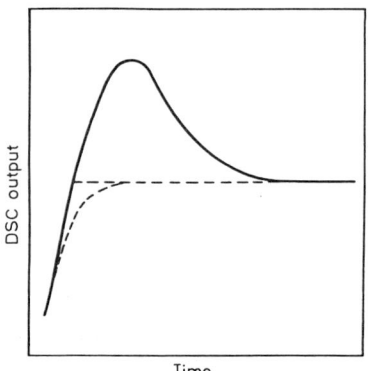

Fig. 4. Representation of DSC output in the isothermal mode for an exothermic reaction, showing baseline construction

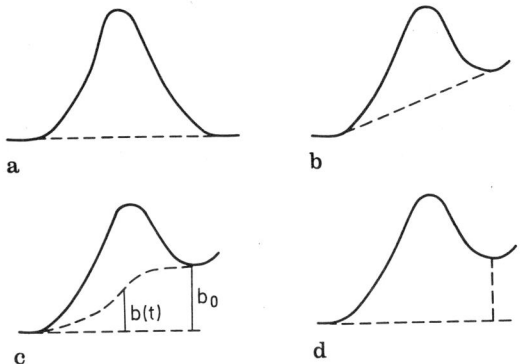

Fig. 5 a–d. Representation of DSC temperature scan for an exothermic reaction, showing baseline construction: a) horizontal; b) linear; c) proportional; and d) linear with vertical boundary

In this case a second DSC scan on the reacted sample may indicate an open-ended peak as in Fig. 5d and a horizontal baseline with an interpolated vertical boundary can be used as an approximation.

2.2.3 Treatment of Kinetic Data

A general expression often used in the analysis of DSC kinetic data is

$$d\alpha/dt = kf(\alpha), \tag{2-6}$$

where at a given time and temperature, $f(\alpha)$ is some function of the fractional conversion, α, and $d\alpha/dt$ is the rate of conversion. The apparent rate constant, k, is usually assumed to be of the Arrhenius form:

$$k = A \exp(-E/RT), \tag{2-7}$$

where A is a constant, R is the gas constant, E is the apparent activation energy, and T is the absolute temperature. For a reaction of order n which involves one reacting substance the conventional kinetic equation is

$$-dc/dt = k_0 c^n \tag{2-8}$$

where c is the concentration of reactant and k_0 is the rate constant for a given temperature. In terms of fractional conversion this gives

$$d\alpha/dt = -(dc/dt)/c_0 = k(1-\alpha)^n \tag{2-9}$$

where c_0 is the initial concentration and k is the apparent rate constant:

$$k = k_0 c_0^{n-1}$$

It is a common practice in thermal analysis literature to quote the apparent rate constant, k, which is convenient although only dimensionally correct for first-order reactions.

A number of studies of simple reactions using DSC have given values for rate constants and activation energy in good agreement with those determined by methods utilising chemical analysis. An example of this is the work of Barrett [18] on the thermal decomposition of free radical initiators.

Many methods have been devised for analysing kinetic data for reactions which can be described by Eq. (2-9). It is convenient to use Eq. (2-9) in its linear logarithmic form

$$\ln r = \ln A + n \ln (1 - \alpha) - E/RT, \qquad (2\text{-}10)$$

where r is the rate, $d\alpha/dt$. For isothermal data the order of reaction may be obtained as the slope of the linear plot of ln r against ln $(1 - \alpha)$, and for experiments at a constant heating rate Eq. (2-10) may be solved numerically by multiple linear regression. Alternatively n may be varied to find the best linear regression fit of $\ln [r/(1 - \alpha)^n]$ on $1/T$. Equation (2-10) can also be used in an incremental form as proposed by Freeman and Carroll [19]:

$$\Delta \ln r = n \Delta \ln (1 - \alpha) - (E/R) \Delta(1/T). \qquad (2\text{-}11)$$

Estimates of n and E may be obtained from a linear plot of $\Delta \ln r$ against $\Delta \ln (1 - \alpha)$ for constant increments of $1/T$.

When the kinetic function is unknown but Eq. (2-6) and (2-7) apply, the apparent activation energy may be determined. From DSC scans at differing heating rates a series of values of reaction rate, r_α, and temperature, T_α, at fixed values of α, can be obtained. The apparent activation energy is then given by [77]

$$\ln r_\alpha = \ln [Af(\alpha)] - E/RT_\alpha \qquad (2\text{-}12)$$

Where $\ln [Af(\alpha)]$ is constant for a given conversion, α. This Eq. can also be applied to isothermal data obtained at different temperatures.

Ellerstein [20] has derived an expression which utilises the second derivative of the conversion with respect to temperature, and assumes the validity of Eq. (2-9). In terms of fractional conversion, and time derivatives the equation may be written as

$$(\dot{r}/r\varphi) T^2 = E/R - nT^2 r/(1 - \alpha) \varphi, \qquad (2\text{-}13)$$

where $r = d\alpha/dt$, $\dot{r} = dr/dt$, and $\varphi = dT/dt$. A linear plot of the left-hand side of Eq. (2-13) against $T^2 r/(1 - \alpha) \varphi$ would yield E/R as the intercept and n as the slope. This method of analysis has been discussed and used by Crane et al [21]. The equation of Rogers and Smith [22] is another form of Eq. (2-13):

$$(1 - \alpha) \dot{r}/r^2 = (1 - \alpha) \varphi E/rT^2 - n \qquad (2\text{-}13a)$$

In this case the data may be plotted in a linear form to yield $-n$ as the intercept and E/R as the slope.

A relationship between the peak maximum temperature, T_m, and E for data obtained at different heating rates was derived by Kissinger [23]:

$$d \ln \varphi / d(1/T_m) = -E/R - 2 T_m . \tag{2-14}$$

This Eq. is however only valid for $n = 1$. A more general relationship at the peak maximum is [22]

$$E/n = RT_m^2 r_m / \varphi (1 - \alpha_m) \tag{2-15}$$

The use of such single-point methods is inadvisable when $n \pm 1$ or the reaction is in other ways complex [22].

Another group of methods for the analysis of data from temperature scans at constant heating rate is based on the integrated rate Eq.

$$\int_0^\alpha d\alpha/f(\alpha) = (A/\varphi) \int_{T_0}^T \exp(-E/RT) \, dT , \tag{2-16}$$

which may be written as

$$F(\alpha) = (A/\varphi) G(T) \tag{2-17}$$

If $f(\alpha) = (1 - \alpha)^n$ and $n = 1$ then

$$F(\alpha) = -\ln (1 - \alpha) \tag{2-18}$$

and for $n \neq 1$

$$F(\alpha) = 1 - (1 - \alpha)^{1-n}/(1 - n) . \tag{2-19}$$

The right-hand side of Eq. (2-17) can be reduced to an exponential integral by substituting $y = E/RT$ to give

$$G(T) = (AE/\varphi R) p(y) \tag{2-20}$$

where

$$p(y) = - \int_{y_0}^y (e^{-y}/y^2) \, dy \tag{2-21}$$

Many methods have been devised for the application of Eqs. (2-16) to (2-20) to thermoanalytical data which involve various approximations of the exponential integral, p(y). Notable examples are the methods of Doyle [24, 25], Horowitz and Metzger [26], Coats and Redfern [27], and Ozawa [28-30]. The method of Ozawa is frequently used. By taking Doyle's approximation for p(y) in Eqs. (2-20) and (2-21) Ozawa obtained the approximate relationship

$$\log \varphi \simeq c - 0.4567 \, E/RT \tag{2-22}$$

where c is constant for a given conversion. The approximation for p(y) is fairly accurate for $20 \leq (E/RT) \leq 50$.

Although temperature scans are experimentally more convenient than isothermal measurements, differences have sometimes been observed between kinetic data from these two methods. In some cases this may be due to the application of inadequate kinetic models in the data analysis, and errors due to thermal lag [31] in the samples. However it has been argued that the isothermal and dynamic rates are intrinsically different [32-34]. Prime [35] has proposed a method of correcting dynamic data, based on these arguments. From the integration of the isothermal rate Eq. (2-6) followed by differentiation with respect to time and temperature, Prime obtained the following expression relating dynamic and isothermal rates

$$(d\alpha/dt)_d = (d\alpha/dt)_i (1 + \varphi Et/RT^2) \qquad (2\text{-}23)$$

where the subscripts d and i refer to dynamic and isothermal conditions, and φ is the heating rate. In some cases this has been applied and the corrected dynamic data were found to be in closer agreement with the isothermal results [36, 37]. Others have argued against the theoretical validity of this approach [38-41]. In general it is advisable to test kinetic results derived from temperature scans against isothermal data [31, 71].

Although the simple rate expressions, Eqs. (2-6) and (2-9), may serve as first approximations they are inadequate for the complete description of the kinetics of many epoxy resin curing reactions. Complex parallel or sequential reactions requiring more than one rate constant may be involved. For example these reactions are often autocatalytic in nature and the rate may become diffusion-controlled as the viscosity of the system increases. If processes of differing heat of reaction are involved, then the deconvolution of the DSC data is difficult and may require information from other analytical techniques. Some approaches to the interpretation of data using more complex kinetic models are discussed in Chapter 4.

3 Epoxy Resin Curing Reactions

Epoxides are notable for their high degree of reactivity towards a variety of nucleophilic and electrophilic reagents. A brief survey of some of the more important reactions relevant to the DSC studies is given here. More extensive and detailed reviews are available elsewhere, see for example Tanaka and Mika [42], and Mika [43]. In general epoxy resin curing reactions involve opening of the epoxide ring followed either by a homopolymerisation reaction with further epoxide, or reaction with other species to form addition products.

Amongst the curing agents of greatest technological importance are the polycarboxylic acid anhydrides, polyamines, and anionic or cationic catalysts. These categories of curing agents are also the ones with which DSC studies of cure have mainly been concerned.

3.1 Carboxylic Acid Anhydride Cure

The use of di- or poly-carboxylic acid anhydrides to cure epoxy resins is based on the reaction of these materials to produce ester links, in the presence of acidic or basic catalysts. Fisch and Hofmann [44, 45] proposed that uncatalysed cure can occur through reactions involving monoester, diester, and ether formation, as in the following scheme.

$$R'OH + OC\overset{O}{\underset{R}{\diamond}}CO \rightarrow R'OOC-R-COOH \qquad (3\text{-}1)$$

$$R'OOC-R-COOH + \overset{O}{CH_2-CH-} \rightarrow R'OOC-R-COOCH_2-\underset{OH}{CH-} \qquad (3\text{-}2)$$

$$R'OH + \overset{O}{CH_2-CH-} \rightarrow R'O-CH_2-\underset{OH}{CH-} \qquad (3\text{-}3)$$

Hydroxylic species (R'OH) are present in oligomers of bisphenol-A diglycidyl ether (BADGE) resins as part of the structure, and may also be present as impurities.

As a result of reaction (3-2), the concentration of hydroxyl will increase and this can produce an autocatalytic effect. In practice basic or acidic catalysts are often added to accelerate the cure, and tertiary amines are frequently used for this purpose.

Overall second-order kinetics have been observed for catalysis by bases, alcohols or acids, and in the base-catalysed reactions the formation of ether groups is relatively insignificant [46, 47]. The base catalysis can be further activated by an acid co-catalyst, HA. For example any resident hydroxyl groups can act as internal co-catalysts. Tanaka and Kakiuchi [48] proposed the following scheme for the reaction catalysed by base (B) with acid co-catalyst:

$$B + HA \rightleftharpoons B \ldots HA \qquad (3\text{-}4)$$

$$B\ldots HA + \overset{O}{CH_2-CH-} \rightleftharpoons \overset{O\ldots HA\ldots B}{CH_2-CH-} \qquad (3\text{-}5)$$
$$\text{(E.HA.B)}$$

$$\text{E.HA.B} + OC\overset{O}{\underset{R}{\diamond}}CO \rightarrow A-OC-R-COOCH_2-\underset{OH}{CH-} + B \qquad (3\text{-}6)$$
$$\text{(HA')}$$

Propagation can proceed with the repetition of steps (3-4) to (3-6) involving the newly formed species, HA', to form a polyester. It was assumed that reaction (3-6) was rate-controlling, and that the concentrations of active species were in a stationary state, which led to a prediction of an overall second-order dependence of rate on anhydride or epoxide concentration.

The possible occurrence of side reactions due to epoxide isomerisation has been considered by Tanaka and Kakiuchi [48] and by Luston et al. [49, 50].

Matejka et al.[51] studied the reaction of phenyl glycidyl ether (PGE) with acetic or benzoic acid anhydride in the presence of benzyldimethylamine and also with benzoic acid as a co-catalyst. They found that the tert-amine is bound irreversibly through the formation of a quarternary ammonium salt; as shown below.

$$R_3N + CH_2-CH(-O-)- \rightarrow -CH(O^-)-CH_2-N^+R_3 \quad (3\text{-}7)$$

$$-CH(O^-)-CH_2-N^+R_3 + R'CO-O-OCR' \rightarrow -C(OOCR')-CH_2-N^+R_3 \cdot R'COO^- \quad (3\text{-}8)$$

It was proposed that propagation proceeds by reaction of carboxyl anion with epoxide to form an ester-alkoxide anion which can react in turn with anhydride to form ester and another carboxyl anion. The rate-determining step was thought to be the reaction of carboxyl anion with epoxide, which would lead to first-order kinetics with respect to epoxide concentration.

Stevens[52,53], using IR spectroscopy, observed that etherification was significant in the reaction of BADGE resins with dicarboxylic acid dianhydrides in the absence of added catalysts. He suggested that the Fisch and Hofmann[44,45] reaction scheme, Eqs. (3-1) to (3-3), can result in second-order kinetics when hydroxyl, anhydride, and monoester are present in approximately equal concentrations, and furthermore that apparent first-order kinetics would apply if hydroxyl or anhydride availability is restricted. The measured rates of reaction of anhydride, esterification, and etherification all showed first-order kinetics for significant extents for a BADGE resin with a high hydroxyl content. More complex behaviour was observed in the region of gelation. Another investigation, by Antoon and Koenig[54] using FTIR spectroscopy, was concerned with BADGE resin cured with methylnadic anhydride (MNA). The observed overall kinetics were first-order up to about 60% conversion, but the data fitted zero- and second-order kinetics almost as well in some cases. The data were interpreted in terms of a reaction scheme in which the first step is the formation of a complex between tert-amine and hydroxyl species in the resin. This was followed by the postulated rate-controlling reaction of the complex with anhydride to form the monoester through a termolecular activated complex. Further reaction of the monoester with epoxide can give the diester and a terminal hydroxyl group, and repetition of this sequence can produce the polyester. By invoking the stationary state hypothesis for active species the model predicts a first-order kinetic dependence on anhydride concentration. However the authors point out that observed changes in concentrations of some chemical species during the reaction do not support the validity of the stationary state hypothesis. Again at higher levels of conversion, close to gelation, the kinetic behaviour was more complex.

In general it is apparent that these reactions are very complex and precise kinetics cannot be predicted with confidence for given compositions and conditions. The early stages of cure may show auto catalytic features while the onset of gelation can introduce a degree of diffusion-control of the kinetics. Orders of reaction between 0 and 4 have been reported, and the apparent order may change during the reaction.

3.2 Amine Addition and Base-catalysed Cure

A great variety of aromatic diamines and aliphatic di- and poly-amines are used as epoxy resin curing agents, and tert-amines can act as catalysts for anionic epoxide homopolymerisation.

The reaction of epoxides with primary or secondary amines involves the following overall reactions[55]

$$RNH_2 + CH_2-\overset{O}{CH}- \rightarrow RNH-CH_2-\underset{OH}{CH}- \qquad (3\text{-}9)$$

$$\underset{|}{RNH} + CH_2-\overset{O}{CH}- \rightarrow \underset{|}{RN}-CH_2-\underset{OH}{CH}- \qquad (3\text{-}10)$$

These reactions are catalysed by acids such as Lewis acids, phenols, and alcohols. The hydroxyl groups formed by the amine epoxide addition are active catalysts, so that the curing reaction usually shows an accelerating rate in its early stages, typical of auto catalysis. In some cases when the amine is present in less than stoichiometric concentrations, reaction of epoxide and hydroxyl may occur to produce an ether group:

$$\underset{|}{RN}-CH_2-\underset{OH}{CH}- + CH_2-\overset{O}{CH}- \rightarrow \underset{|}{RN}-CH_2-\underset{O-CH_2-\underset{OH}{CH}-}{CH}- \qquad (3\text{-}11)$$

A study of the reaction of PGE and aniline using high purity reagents and excluding traces of hydroxylic materials, has been reported by Enikolopiyan [56]. This demonstrated that amine-epoxide addition can occur in the absence of acidic catalysts, and the observed overall rate was second-order in amine concentration.

Schechter [55] proposed that the catalytic effect of hydroxyl groups on the epoxide-amine addition reaction involved a termolecular activated complex formed in the concerted reaction of amine, epoxide and hydroxyl. Smith [57] suggested a modified mechanism in which the same activated complex is formed in a bimolecular reaction between an adduct formed from epoxide (E) and the proton donor (HX), and the amine:

$$E + HX \rightleftharpoons E \ldots HX \qquad (3\text{-}12)$$

$$\underset{|}{RNH} + E \ldots HX \rightleftharpoons \underset{|}{RNH} \ldots \overset{O\ldots HX}{CH_2-CH-} \qquad (3\text{-}13)$$
$$(C)$$

$$C \rightarrow \underset{|}{-N}-CH_2-\underset{OH}{CH}- + HX \qquad (3\text{-}14)$$

The rate-determining step is the reaction of the activated complex, Eq. (3-14).

Tanaka and Mika [42] suggest that the higher basicity of amine relative to epoxide makes the formation of an amine-proton donor adduct more likely, and they proposed the following equations as an alternative to Eqs. (3-12) and (3-13).

$$-\overset{|}{N}H + HX \rightleftharpoons -\overset{|}{N}H \ldots HX \qquad (3\text{-}15)$$

$$-\overset{|}{N}H \ldots HX + E \rightleftharpoons -NH \ldots CH_2-CH- \qquad (3\text{-}16)$$
$$ \therefore XH \therefore O$$

The kinetic effects of different relative reactivities of hydroxyl and amine on the mechanisms of Smith, and of Tanaka and Mika has been considered by King and Bell [58], who offer some experimental evidence in favour of the Tanaka and Mika mechanism. As well as showing auto catalysis in the early stages of cure, these reactions often exhibit a pronounced retardation in their later stages. This is often ascribed to diffusion effects as the viscosity increases following gelation of the system, but chemical effects may also be important, for example it has been suggested [56] that non-reactive complexes may be produced from hydroxyl groups and sec- or tert-amines.

In practice the amine curing reactions are often accelerated by the addition of Lewis acids, especially amine complexes of boron trifluoride. Such materials can also initiate epoxide homopolymerisation in which chain propagation occurs through a carbocation:

$$Y^+ + CH_2-CH- \xrightarrow{} Y-O-CH-CH_2^+$$
$$ \xrightarrow{} Y-[O-CH-CH_2]_x-O-CH-CH_2^+ \qquad (3\text{-}17)$$

An analogous homopolymerisation can be initiated by strong bases, including for example tert-amines. In this case chain propagation probably proceeds through an oxyanion:

$$X^- + CH_2-CH- \xrightarrow{} X-CH_2-CH-O^-$$
$$ \xrightarrow{} X-[CH_2-CH-O]_x-CH_2-CH-O^- \qquad (3\text{-}18)$$

Most of the reported DSC studies involving base-catalysed cure are concerned with dicyanodiamide (DICY); which is a crystalline material, melting at 208 °C. At ambient temperature the solubility in epoxy resins is very low but above about 100 °C it is sufficiently soluble to initiate cure. DICY has free amino-hydrogen groups which can partake in addition reactions with epoxides:

$$4\ CH_2-CH- + H_2N-C=NH \rightarrow (-CH(OH)CH_2)_2N-C=N-CH_2-CH(OH)-$$
$$ NH-CN N-CN$$
$$ CH_2-CH(OH)- \qquad (3\text{-}19)$$

Concentrations of DICY below the stoichiometric level are normally used so that part of the cure is dependent on its ability to promote homopolymerisation and epoxide-hydroxyl reaction. The tert-amine-catalysed reaction of DICY and PGE was studied by Saunders et al. [59], using IR and NMR spectroscopy. At 93 °C the DICY slowly dissolved and reacted mainly to produce N-alkylcyanoguanidines, as in reaction (3-19), together with the occurrence of some anionic epoxide homopolymerisation. At temperatures above 100 °C there was a slower addition reaction of secondary hydroxyl and nitrile groups to form an iminoether with subsequent rearrangement to a guanylurea structure. Additional complications in the reaction of BADGE resin and DICY were reported by Sacher [60] who observed the formation of melamine from DICY, and found that the rate of cure was dependent on the DICY particle size. It has been suggested by Schneider [61] et al. that polyether formation from epoxide-hydroxyl addition reactions dominates the later stages of cure.

Substituted urea compounds are often used as accelerators for DICY cure, specifically 3-(4-chlorophenyl)-1,1-dimethylurea (Monuron) and the 3-(3,4-dichlorophenyl) derivative (Diuron). This introduces a further level of complexity to the curing mechanism. Son and Weber [62] showed that Monuron could dissociate to form dimethylamine and 4-chlorophenylisocyanate and that this is facilitated by the reaction of DICY with the isocyanate to form a carbanilinoguanidine:

$$Cl-C_6H_4-NHCON(CH_3)_2 \rightleftharpoons Cl-C_6H_4-NCO + (CH_3)_2NH \quad (3\text{-}20)$$

$$Cl-C_6H_4-NCO + DICY \longrightarrow Cl-C_6H_4-NHCONH-\underset{\underset{CN}{\overset{|}{NH}}}{C}=NH \quad (3\text{-}21)$$

It was suggested that the dimethylamine would react with epoxide and that the resulting tert-amine would catalyse epoxide homopolymerisation. LaLiberte et al. [63] found that Monuron or Diuron could react with epoxide to form an oxazolidone and dimethylamine:

$$-\overset{O}{\overset{\diagup \diagdown}{CH-CH_2}} + RNHCON(CH_3)_2 \longrightarrow -\underset{\underset{CH_2-NR}{|}}{\overset{O}{\overset{\diagup \diagdown}{CH \quad CO}}} + (CH_3)_2NH \quad (3\text{-}22)$$

Dimethylamine was shown to have a strong accelerating effect on the cure of a BADGE/DICY mixture at 116 °C and it was also found that Monuron undergoes considerable reaction with the resin at 90 °C.

Another group of strong bases used as epoxy resin curing agents are the imidazoles:

(imidazole structure with substituents R_1, R_2, R_3)

These compounds can initiate anionic polymerisation of epoxides, and when $R_1 = H$ the secondary amine can react by addition to an epoxide group. Farkas and Strohm [64] have studied the reaction of 2-ethyl-4-methyl imidazole with phenyl glycidyl ether and BADGE resin using chemical analysis and proton NMR spectroscopy. They found that the imidazole readily forms adducts with epoxide of 1:1 and 1:2 molecular ratio:

$$\text{(3-23)}$$

$$\text{(3-24)}$$

They also found that the adducts were good catalysts for curing BADGE resin and proposed that they are the catalytic species formed in the first stage of imidazole-initiated polymerisation.

4 Results of DSC Studies

4.1 Anhydride Curing Agents

The first notable report on the application of DSC to the study of epoxy resin cure was by Fava [65]. The resin system was bisphenol-A diglycidylether (BADGE) with hexahydrophthalic anhydride (HHPA) as the curing agent and tris-2,4,6-dimethyl aminomethyl phenol as an accelerator, in the weight proportions 100/80/1. DSC experiments were in the isothermal and temperature-scanning (dynamic) modes, on the uncured resin. In some cases after partial isothermal cure, temperature scans were run in order to measure the glass transition temperature (T_g) and residual exotherm. An ingenious method was used to attempt to calculate isothermal conversion data from the results of a set of DSC scans at different heating rates, and this is depicted in Fig. 6a. The reciprocal rate of cure is plotted against the fractional extent of cure in Fig. 6b. The partial area under such a curve is the time to reach a given degree of cure at the reference temperature, T_0:

$$\int_{\alpha_0}^{\alpha} (dt/d\alpha)\, d\alpha = t_{\alpha-\alpha_0} \qquad (4\text{-}1)$$

A practical problem is that the curve is asymptotic at $\alpha = 0$ so that data can only be obtained after a significant initial degree of reaction. The results obtained by this method are shown in Fig. 7, and they are in fairly good agreement only in the temperature range 390–420 K. The curves calculated from temperature scans predict iso-

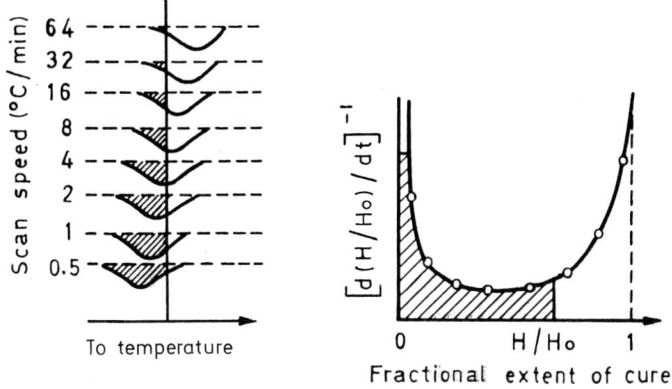

Fig. 6. DSC results for epoxy resin cure with a carboxylic acid anhydride: a) Set of displaced thermograms; b) Curve deduced from a) and used to obtain isothermal cure data at temperature T. (From Ref. [65], Fig. 3)

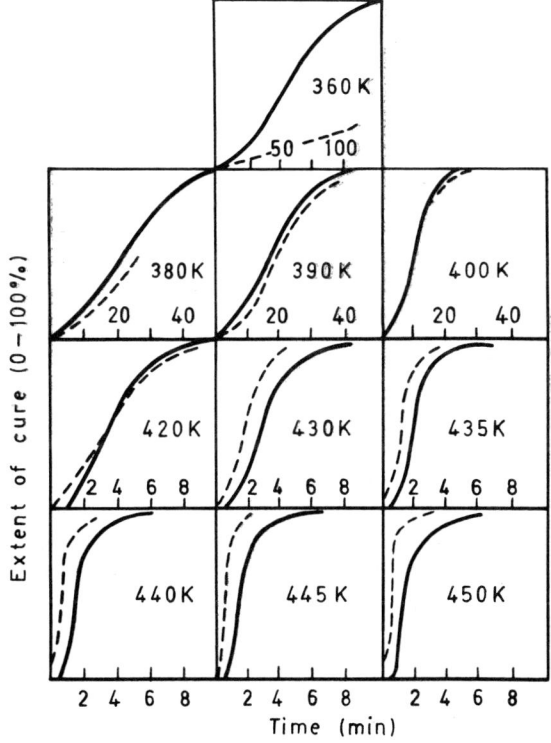

Fig. 7. Comparison of cure curves at various temperatures, obtained from isothermal measurements (solid curves), and calculated from DSC scans at different heating rates (dashed curves) (From Ref. [65], Fig. 4)

thermal rates which are too low at lower temperatures and too high at the higher temperatures. Fava concludes that at the lower temperatures the temperature scans do not provide sufficient data for reliable isothermal predictions, while the isothermal method is inaccurate at the higher temperatures because a significant degree of reaction is occurring before the establishment of temperature equilibrium at the start of the experiment. The use of the true isothermal baseline, as discussed in 2.2.2.2, should improve the accuracy of the direct isothermal method. A plot of log maximum rate against 1/T was linear and yielded an apparent activation energy of 74 kJ/mole. This implies that the conversion at maximum rate is constant and that Eq. (2-12) applies to this system.

Other DSC studies have been made on the same type of resin (BADGE) in its various commercial grades [66–70]. Peyser and Bascom [66] used BADGE resin with methyl-bicyclo-[2,2,1]heptane-2,3, dicarboxylic anhydride, known as methylnadic anhydride (MNA), as the curing agent with benzyldimethylamine (BDMA) as an accelerator, in the weight proportions 100/87.5/1.5. DSC scans were made at various heating rates on samples in unsealed pans. At a heating rate of 10 K/min a significant amount of sample volatilisation was observed, but this was greatly reduced at lower heating rates. The heat of cure was found to decrease with increasing heating rate, and this was attributed to reduced volatilisation. The baselines under the DSC scans were constructed so as to be proportional to the degree of conversion by an iterative method as discussed in 2.2.2.2. It was assumed that the kinetics could be described by Eq. (2-9) and the data were analysed using Eqs. (2-10), (2-14), and the Freeman-Carrol method, Eq. (2-11). Plots of $\ln r - n \ln (1 - \alpha)$ against 1/T, according to Eq. (2-10), for $n = 2$, gave two linear regions. Up to about 12% conversion the apparent activation energy, E, was about 63 kJ/mole, and at higher conversions E was about 159 kJ/mole. This increase in E was attributed to the onset of diffusion control after about 12% conversion. These authors [67] also studied the cure of the same resin and accelerator with hexahydrophthalic anhydride (HHPA) using stoichiometric proportions of epoxide and anhydride. In this case no significant volatilisation of the sample was observed during the experiments. From both DSC scans and isothermal experiments the results gave a good fit to Eq. (2-10) for the order of reaction, n, in the range 0.8–1.2, and $E \simeq 105$ kJ per mole, although the isothermal data indicate autoacceleration in the early stages of reaction. The results were interpreted in terms of the stationary state mechanism of Tanaka and Kakiuchi [48] with reaction (3-5) as the rate-controlling step. The observed first-order kinetics would also be consistent withe the mechanism of Matejka et al. [51].

Möhler and Schwab [68] studied the cure of BADGE resin with various anhydrides and accelerators, in the presence of a quartz powder filler, using DSC at only one heating rate, 5 K/min. Heat flow data were obtained by assuming linear baselines under the DSC exotherms and the data were analysed by the Freeman-Carroll method, Eq. (2-11). The constants n and E were obtained from regions approximating to linearity in the plots of $\Delta \ln (d\alpha/dT)/\Delta \ln (1 - \alpha)$ against $\Delta(1/T)/[-\Delta \ln (1 - \alpha)]$. In most cases two linear regions were observed in the plots. The change in slopes at higher conversions indicated an apparent increase in both n and E which was attributed to an onset of diffusion control.

Malavasic et al. [69] examined the system consisting of BADGE resin (Araldite CY 206), a carboxylic acid anhydride with acid number 319, of unspecified composition, and BDMA in the weight ratio 100/85/2. From isothermal measurements in the range 370–380 K, the rate data were fitted to the integrated forms of Eq. (2-9) for n = 0.5 or 1. In the conversion range 18–20% an approximate fit was obtained for n = 0.5, and above 80%, for n = 1. The apparent activation energy was 56–58 kJ per mole.

An epoxy potting compound consisting of a BADGE-type resin with an unspecified anhydride is the subject of a study by Swarin and Wims [31], using both isothermal and dynamic DSC. The data were analysed using kinetic equations in the form of Eq. (2-9) and (2-10). The criterion of best fit was the linearity of Arrhenius plots for different values of the reaction order, n, and the best fits were obtained for n = 1. The values obtained for the first-order rate constant from dynamic experiments were found to decrease with decreasing heating rate, and this was ascribed to the effect of thermal lag in the samples. An extrapolation of the apparent rate constants to zero heating rate was used to obtain quasi-isothermal values. Isothermal conversion-time curves were calculated from the extrapolated dynamic data and these were in good agreement with experimental isothermal data at 110, 135, and 160 °C.

A different approach was taken by Stevens and Richardson [70] who used the change in T_g with degree of cure as a kinetic monitor. These authors worked with BADGE resins (Araldite CT 200 and CY 207) and either phthalic anhydride (HT 901) or a phthalic/tetrahydrophthalic anhydride mixture (HT 903) as the curing agent, without added catalyst. The mass ratios of the two systems were CT 200/HT 901, 100/30, and CY 207/HT 903, 100/60 which correspond to epoxide/anhydride molar ratios of 1/0.85 and 1/0.91, respectively. Resin CT 200 has lower epoxide and higher hydroxyl content than CY 207. Approximate overall first-order kinetics were found at up to 85–92% conversion. Stevens [52] also made a study of these systems using IR spectroscopy to monitor changes in anhydride, aromatic ester, and ether group concentrations during cure. The results of the two methods are compared in Table 1.

It appears that the rate of change in T_g, associated with an increasing degree of polymerisation and crosslinking due to cure, is close to the rate of etherification.

The results of DSC studies on the anhydride cure of epoxy resins are summarised in Table 2. These studies have confirmed that the cure mechanism is complex. The early stages show autocatalytic features while the later stages are complicated by the effects of diffusion control. Intermediate stages of cure can show an approximation to overall kinetic orders of 1 or 2. In general the isothermal DSC data are easier to

Table 1. First-order rate constants for the cure of CT 200/HT 901 and CY 207/HT 903 from DSC and IR (Ref. [70])

Method	10^5 k (s^{-1})	
	CT 200/HT 901 125 °C	CY 207/HT 903 120 °C
DSC (T_g)	4.2	1.1
IR (Ether)	5	1.1
IR (Ester)	6.5	4
IR (Anhydride)	8.6	2.9

Table 2. Summary of DSC kinetic results on anhydride cure

Resin (BADGE)	Anhydride	Catalyst	Isothermal (I) or Dynamic (D) method	n^a	$\ln A^a$ (s^{-1})	E kJ/mole	Note	Ref.
DER332LC	HHPA[b]	TAP[c]	I (84-182 C)	—	—	74		65)
DER332	MNA[d]	BDMA[e]	D	2	11	63	<12% conv.	66)
			D	2	42	159	12-90% conv.	
DER332	HHPA	BDMA	I (125-157 C)	1	27	105	10-90% conv.	67)
			D					
CY205	HHPA	BDMA	D	0.5	23	97	5-22% conv.	68)
				1.0	26	111	>22%	
CY205	HHPA	MI[f]	D	0.5	22	95	2-22% conv.	68)
				1.6	34	136	>22%	
CY205	MTPA[g]	BDMA	D	1.1	26	101	>2% conv.	68)
CY205	MTPA	MI	D	1.0	25	107	3-18% conv.	68)
				1.4	26	111	>18%	
CY205	DSA[h]	BDMA	D	0.2	15	74	3-14% conv.	68)
				0.8	19	90	>14%	
CY205	DSA	MI	D	1.0	19	85	2-35% conv.	68)
				1.4	29	121	>35%	
CY206	EXP[i]	BDMA	I (99-108 C)	0.5	16	58	18-80% conv.	69)
				1.0	12	56	>80%	
CT200	PA[j]	—	Tg[l]	1.0	($k = 4.2 \times 10^{-5}$ s^{-1} at 125 C)		<85% conv.	70)
CY207	PA/THPA[k]	—	Tg	1.0	($k = 1.1 \times 10^{-5}$ s^{-1} at 120 C)		<85%	70)
m	m	m	n	1.0	25	98		31)

[a] Eq. (2-7) and (2-9);
[b] Hexahydrophthalic anhydride;
[c] Tris-2,4,6-dimethylaminomethylphenol;
[d] Methylnadic anhydride;
[e] Benzyldimethylamine;
[f] 1-Methylimidazole;
[g] Methyltetrahydrophthalic anhydride;
[h] Dodecylsuccinic anhydride;
[i] Experimental anhydride, acid no. 392;
[j] Phthalic anhydride;
[k] Phthalic/Tetrahydrophthalic anhydride mixture;
[l] From change in Tg;
[m] Not specified;
[n] Data extrapolated to zero heating rate

4.2 Primary and Secondary Amines as Curing Agents

In general the amine-epoxy resin curing reactions show complex kinetics typified by an initial acceleration due to autocatalysis, while the later post-gelation stages may exhibit retardation as the mechanism becomes diffusion-controlled. However some workers [72-80] have found that over a limited range of conversion the kinetic data may be described by the simple models of Eq. (2-6) or (2-9).

Prime [73] studied the reaction of BADGE resin with metaphenylenediamine (MPD) using isothermal and dynamic DSC. The data were fitted to Eq. (2-9) and to this Eq. modified to correct for heating rate (see 2.2.3). Use was also made of the Kissinger equation, Eq. (2-14), relating the order of reaction and the activation energy to data measured at the exotherm peak maximum. Further isothermal studies were made on the same resin system [74, 76] using DSC, IR spectroscopy and dc conductivity methods, which yielded kinetic data in good agreement. Prime and Sacher [75] examined the cure of BADGE resin with a polyamide-amine mixture, Versamid 140, as the curing agent. DSC scans gave complex exotherms which could be resolved into three component peaks. The first of these resolved peaks accounted for 79 % of the total exotherm area and it was attributed to the reaction of epoxide with primary and secondary amine. In this and a further study [76] the kinetic data for the early part of the reaction were fitted to Eq. (2-9).

In an examination of the reaction of BADGE resin with 4,4'-diaminodiphenylmethane (DDM) Barton [77] used the general equation, Eq. (2-6), to determine apparent activation energy, E, from dynamic and isothermal DSC data. If it is assumed that $f(\alpha)$ in Eq. (2-6) does not depend on temperature then for isothermal data, E may be found from the slope of a linear plot of $\ln r_\alpha$ against $1/T$, where r_α is the rate at conversion α from experiments at different temperatures. While for dynamic experiments data may be taken from scans at different heating rates to give r_α, T_α at a given conversion for evaluation in the Arrhenius plot corresponding to Eq. (2-12). At 50 % conversion the isothermal and dynamic data gave similar values for E; 49.6 and 47.0 kJ/mole, respectively. The validity of Eq. (2-6) and (2-12) implies that plots of isothermal conversion against log time should be superimposable by shifts along the log time axis. This is illustrated in Fig. 8, and it is seen that up to about 50 % conversion, the data fit the shifted 158 °C curve fairly well. The log time shifts used in constructing Fig. 8 correspond to an activation energy of 51.1 kJ/mole, in quite good agreement with the value obtained from Eq. (2-12). In a further study [78] on the same system, isothermal conversion as a function of time at temperatures between 140 and 180 °C, was calculated using data from a DSC scan at 10 K/min heating rate. The scan data were first reduced to a fixed temperature using the experimentally determined activation energy for a given degree of conversion. The reduced rate data were then transformed to conversion-time data by Fava's [65] method which utilizes the relationship of Eq. (4-1). The time values are found by numerical integration of the reciprocal rate data as a function of α. Experimental and calculated results are shown in Fig. 9. The discrepancies at higher conversions are likely to be associated with

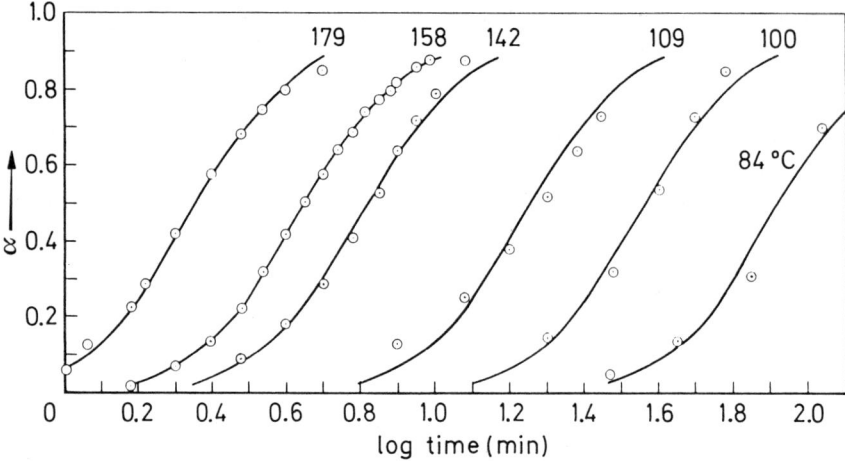

Fig. 8. Plots of fractional conversion against log time from isothermal experiments for BADGE/DDM. Points are experimental; solid curves are the 185 °C curve shifted horizontally to superimpose on experimental points (From Ref. [77], Fig. 4)

diffusion-controlled processes not accounted for in the simple kinetic model. This conclusion is supported by isothermal studies which show that at temperatures below the glass transition temperature (T_g) of the fully cured resin the reaction virtually ceases as the T_g approaches the cure temperature.

Cizmecioglu and Gupta [79, 81] worked with a commercial resin formulation, NARMCO 5208, which they found to consist of the resin, tetra-N-glycidyl-4,4'-diamindiphenylmethane (TGDDM), and the curing agent 4,4'-diamindiphenylsulphone (DDS), together with an epoxy novolac resin in the weight ratio 100/28/8.2. DSC scans gave a single cure exotherm peak and the data from single scans were analysed by the method of Ellerstein [20], Eq. (2-13). The activation energy was also determined by the Kissinger method, Eq. (2-14) from the effect of heating rate on

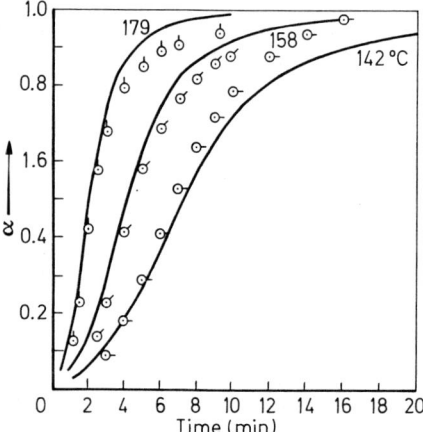

Fig. 9. Isothermal cure curves predicted from dynamic data, and corresponding isothermal experimental points (From Ref. [78], Fig. 3)

the exotherm peak temperature. The apparent order of reaction was in the range 1.3–2.0 and appeared to decrease with increasing heating rate, while E was 109 ± 8 kJ per mole from both methods of determination.

The reactions of BADGE, the diglycidyl ether of 1,4-butanediol (BDDGE), and 1,2,7,8-diepoxyoctane (DEPO) with tetraethylenepentamine (TEPA) have been studied by Severini et al. [82] using dynamic DSC at 20 K/min heating rate. Also for BADGE/TEPA and BDDGE/TEPA isothermal measurements were made at 77 to 97 °C. The activation energy for all four resins, found from the dynamic data by the Freeman-Carrol method, Eq. (2-11), was about 99 kJ/mole, whereas from the isothermal data using Eq. (2-12) it was about 41–43 kJ/mole. Such a large difference between activation energy determined by different methods arouses doubts about the adequacy of the simple kinetic model used in the analysis.

The work described above has utilised kinetic models involving a single rate constant, but this cannot describe the observed autocatalytic nature of the early stages of the curing process. A notable publication on epoxy resin-amine cure kinetics which is concerned with this autocatalysis is the paper by Horie, Hiura, Sawada, Mita and Kambe [83]. This work involved a DSC investigation, in the isothermal mode (50–70 °C) of the reaction of epoxides with aliphatic amines. The reaction of phenylglycidyl ether (PGE) with n-butylamine (BA) was used as a model system, and the study was extended to the curing reaction of BADGE resin with several aliphatic diamines. The heat of reaction of PGE with BA was determined by isothermal DSC at 70 °C, for mixtures with a mole fraction of BA between 0.2 and 0.7. The heat of reaction was found to be 102.5 ± 2.5 kJ/mole epoxide. The heat of reaction of PGE and BADGE resin with a variety of curing agents has also been determined by Klute and Viehmann [84]. For primary amines the value was about 109 kJ/mole, in fairly good agreement with the data of Horie et al. Tertiary amines and the boron trifluoride-diethyl ether complex, which can initiate epoxide homopolymerisation, gave a lower heat of reaction, about 92 kJ/mole. Isothermal DSC data obtained for the BADGE/BA reaction at 70 °C are shown in Fig. 10, which illustrates the initial autoacceleration typical of these reactions.

The interpretation of the kinetics is based on the mechanisms proposed by Schechter et al. [55] and Smith [57], for the reaction of secondary amines with epoxides, extended to include the primary amine reaction. The rate-determining step is assumed to be the reaction of amine, epoxide and hydroxyl or other proton-donor species, HX, to form a termolecular complex, Eq. (3-13). The proposed reaction scheme is:

$$A_1 + E + HX_A \xrightarrow{k_1} A_2 + HX_A \tag{4-2}$$

$$A_1 + E + HX_0 \xrightarrow{k_1'} A_2 + HX_0 \tag{4-3}$$

$$A_2 + E + HX_A \xrightarrow{k_2} A_3 + HX_A \tag{4-4}$$

$$A_2 + E + HX_0 \xrightarrow{k_2'} A_3 + HX_0 \tag{4-5}$$

where A_1, A_2, and A_3 are primary, secondary, and tertiary amine, E is epoxide, HX_0 is hydroxyl resident in the system as impurity or additives, and HX_A is hydroxyl

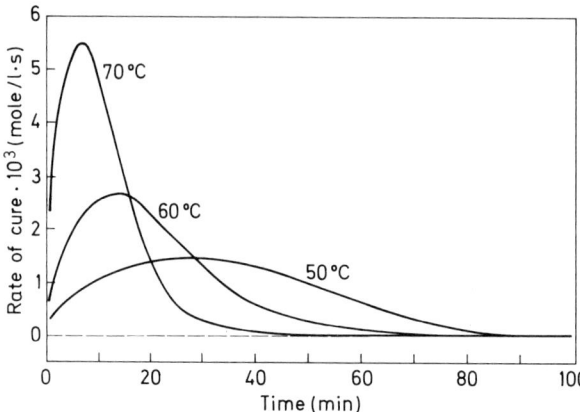

Fig. 10. DSC curves for isothermal cure of BADGE with EDA (From Ref. [83], Fig. 6)

groups formed in the amine-epoxide addition reaction. It has been reported [55] that the reaction between ethylamine and PGE occurs without any etherification so that HX is not consumed and may act as a true catalyst. From this scheme an overall kinetic Eq. was derived:

$$dx/dt = (k'_1 c_0 + k_1 x)(e_0 - x)(a_1 + a_2) \qquad (4\text{-}6)$$

where X is the epoxide consumed at a given time, c_0 and e_0 are the initial concentrations of HX_0 and epoxide, a_1 and a_2 are the concentrations of primary and secondary amine, and $n = k_2/k_1 = k'_2/k'_1$, the relative reaction rate of secondary and primary amine with epoxide. An equivalent expression in terms of initial concentration of primary amine, a_0, is

$$dx/dt = (k'_1 + k_1 x)(e_0 - x)(a_0 - x/2) f \qquad (4\text{-}7)$$

where f is given by

$$f = 1 + a_2 \Delta n/(a_1 + a_2/2) \qquad (4\text{-}8)$$

and

$$n = 0.5 + \Delta n \qquad (4\text{-}9)$$

When the primary and secondary amines are of approximately equal reactivity, $\Delta n \simeq 0$ and $f \simeq 1$. Eq. (4-7) may be written in terms of the fractional conversion, α, and this form for $\Delta n = 0$ is

$$d\alpha/dt = (K_1 + K_2 \alpha)(1 - \alpha)(1 - R\alpha) \qquad (4\text{-}10)$$

where $K_1 = k'_1 a_0 c_0$, $K_2 = k_1 a_0 c_0$ and $R = e_0/2a_0$. The data for the PGE/BA reaction were fitted to Eq. (4-7) for $f = 1$. Plots of the reduced rate, $(dx/dt)/(e_0 - x)(a_0 - x/2)$ against X were linear up to about 50% conversion. An increase in slope at higher conversions was attributed to the error in assuming that n = 0.5, and the true value of n was estimated to be in the range 0.6–0.7. The addition of n-butanol to the system had the predicted effect of increasing the initial rate and shifting the exotherm peak to earlier times. At 70 °C the rate constants k_1, k_2, k_1', k_2' were 1.9, 1.2, 1.5, and 0.9 10^4 l^2 mole^{-2} s^{-1}, respectively. Arrhenius plots for k_1 and K_2 gave activation energies of 58 and 56 kJ/mole.

The same method was applied to the reaction of BADGE resin with ethylenediamine (EDA), trimethylenediamine (TMDA) and hexamethylenediamine (HMDA). The reduced rate plots are shown in Fig. 11. In each case after a linear region up to about 60–70% conversion, a pronounced retardation in rate was observed, and this was attributed to the reaction becoming diffusion-controlled at the onset of gelation. This onset of diffusion control occurred at increasing conversion with increasing length of the diamine aliphatic chain. The activation energy for the reaction was in the range 54–57 kJ/mole for all three diamines. Further evidence for diffusion control is that the final conversion at a given temperature was limited as the glass transition temperature of the network approached the cure temperature. For example at 50 °C the final conversion was about 70% for the BADGE/EDA system. It was assumed that in the later stages of reaction the increase in T_g relative to that of linear polymer was proportional to the increase in cross-link density. The cross-link density could be calculated as a function of conversion from the stoichiometry and the relative reactivity of primary and secondary amine. Good agreement was found between the calculated T_g and the observed maximum conversion at each cure temperature.

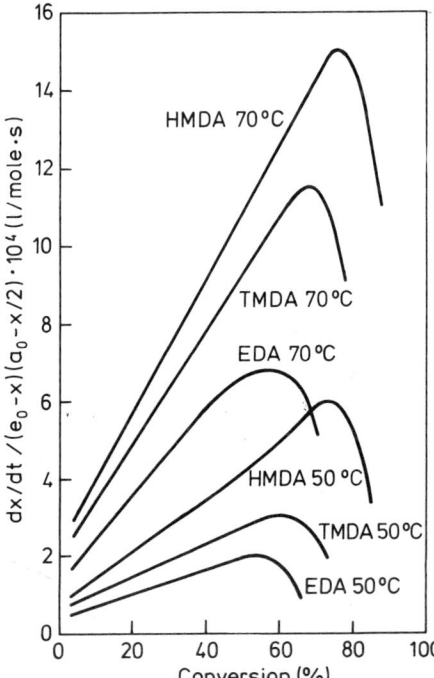

Fig. 11. Reduced rate curves for isothermal cure of BADGE with EDA, TMDA, and HMDA at 50 and 70 °C (From Ref. [83], Fig. 8)

The change in T_g in the later stages of cure can provide a sensitive index of the extent of cure. For example in the BADGE/DDM system [78] it was found that the final 10% of conversion corresponded to a 70 K increase in T_g.

The cure of the BADGE/EDA stoichiometric mixture has also been studied by Ricciardi et al. [115] using isothermal DSC at 40–80 °C, and temperature scans with a presure DSC cell at 4.5×10^6 Pa, and various heating rates, which gave exotherms in the range 40–200 °C. The isothermal data obtained at ambient pressure were analysed by the method of Horie et al. to give plots of $(d\alpha/dt)/(1 - \alpha)^2$ against α which were linear up to about 60% conversion. The rate constants derived for the hydroxyl-catalysed reaction were in fair agreement with those of Horie et al. The dynamic data followed overall second-order kinetics (Eq. (2-9) with n = 2). From these dynamic data the apparent pre-exponential factor, A, was 3.5×10^{12} s^{-1}, and the activation energy was 102.6 kJ/mole. It was proposed that at the higher temperatures of the DSC scans the reaction is controlled by the uncatalysed amine-epoxide addition, while the lower temperature isothermal cure involves this process together with the hydroxyl-catalysed reaction.

Various isothermal DSC studies on epoxy-amine cure have been reported in which the kinetic data are fitted to a generalised form of Equation (4-10):

$$r = d\alpha/dt = (K_1 + K_2\alpha^m)(1 - \alpha)^n \qquad (4\text{-}11)$$

Kamal et al. [85] examined the stoichiometric mixture of BADGE and MPD in the temperature range 57–147 °C. The data were fitted to Equation (4-11) with m = n = 1. In a later study on the same system Sourour and Kamal [86] used amine to epoxide equivalent ratios (B) of 1.0 and 1.5 and found that the data fitted the Eq.

$$r = (K_1' + K_1'\alpha)(1 - \alpha)(B - \alpha) \qquad (4\text{-}12)$$

which is equivalent to Eq. (4-10) whith B = 1/R, $K_1' = K_1 B$ and $K_2' = K_2 B$. Plots of reduced rate against α were linear up to about 40–50% conversion, followed by a marked ratardation in reduced rate, similar to the effect observed by Horie et al. for aliphatic amines. The activation energy associated with K_1' was about 84 kJ/mole and about 46 kJ/mole for K_2'. The critical conversion at gelation was calculated from Flory-Stockmayer statistics to be 58% for B = 1 and 78% for B = 1.5. Good agreement was found between viscometric gel times and DSC measurements of the time to reach the calculated critical conversion, while the activation energy for gelation was about 54 kJ/mole.

The BADGE/MPD system has also been examined by Ryan and Dutta [87] using Eq. (4-11). It was assumed that m + n = 2, and a numerical solution of Eq. (4-11) for the special case at maximum rate ($d^2\alpha/dt^2 = 0$) gave an apparent linear decrease in m with increasing temperature from about 1.2 at 60 °C to 0.6 at 177 °C. A similar approach was taken by Foun et al. [116] who studied the resins BADGE and butanedioldiglycidylether with stoichiometric proportions of DDM or MPD. Isothermal DSC kinetic data in the range 100–140 °C were fitted to equation (4-11) with m + n = 2. The results for BADGE resin are included in Table 3. Flammersheim et al. [88] studied the cure of BADGE with N,N'-dibenzylethylenediamine and with benzylamine. The data gave a good fit to Eq. (4-11) for m $\simeq$ 1 and n $\simeq$ 1.5 over the temperature range 70–105 °C. Reduced rate-conversion plots were linear up to about 90% conversion with no evidence of the usual deviation due to diffusion control.

The cure of tetra-N-glycidyl-4,4'-diaminodiphenylmethane (TGDDM), as a commercial grade, with 4,4'-diaminodiphenylsulphone (DDS) has been studied by Barton [89], in the temperature range 170–220 °C. The data were fitted to Eq. (4-11) with m = 1 and n = 1 or 2.

Better fits were obtained for n = 1 which gave linear reduced rate-conversion plots up to 20–30% conversion, followed by a downward curvature. The apparent pre-exponential factors and activation energies associated with K_1 and K_2 were $A_1 = 6.53 \times 10^5 \, s^{-1}$, $E_1 = 80.4 \, kJ/mole$, $A_2 = 3.01 \times 10^5 \, s^{-1}$, and $E_2 = 71.3 \, kJ/mole$. These kinetics can be explained in terms of a bimolecular rate-determining step between hydroxylic catalyst species and either amine or a rapidly-formed amine-epoxide adduct. An analysis similar to that of Horie et al. yields the kinetic Eq.

$$d\alpha/dt = (K_1 + K_2\alpha)(B - \alpha) f \qquad (4-13)$$

where the parameters are defined as previously, but with reference to different rate-controlling reactions, and with the condition that $f \simeq 1$ when $n \simeq 0.5$. The mechanism may be dependent on the relative basicity of the amine, and in this respect DDS is deactivated by the sulphone group.

The relative reactivity of secondary and primary amines with epoxide, k_2/k_1, can affect the overall kinetics. If k_2/k_1 is not close to 0.5 the parameter f in Eqs. (4-7) and (4-13) becomes significant. In this respect there are differences between aliphatic and aromatic or alicyclic amines. For the aliphatic amines the ratio k_2/k_1 is reported to be in the range 0.6–0.7 [83, 90], whereas for aromatic amines values in the range 0.2 to 0.5 have been observed [90–96]. For the reaction of BADGE with DDS, Dobas et al. [94] reported a value of 0.21 for k_2/k_1 at 80 °C. This factor is likely to account for part of the observed non-linearity in reduced rate plots at higher levels of conversion.

Another aspect that has not been taken into account in the kinetic models discussed so far is the occurrence of ether-forming reaction through epoxide homopolymerisation or reaction with hydroxyl groups. In the system TGDDM with an initial DDM concentration less than the stoichiometric level the overall conversion of epoxide is greater than that expected for epoxide-amine addition [89, 97, 98].

Gupta et al. [99] studied the reaction of TGDDM with DDS and di-N-methyl-DDS (DMDDS) using DSC and spectroscopic methods. The DSC scans indicated an overall heat of reaction, Q_0, of 99 kJ/mole epoxide for the mixture containing 22 wt.-% DDS, and 75 kJ/mole for 22 wt.-% di-N-methyl-DDS. The quoted standard deviations for Q_0 were 6.4 and 9.8, respectively. It was concluded that the differences in Q_0 were not significant, that the epoxide-secondary amine reaction was negligible, and that the reaction in the presence of secondary amine was mainly etherification. There is however a conflicting observation [89] that TGDDM/DDS mixtures with amine/epoxide equivalent ratios in the range 0.61 to 1.14 have a constant heat of reaction of about 110 kJ/mole epoxide. This can be interpreted by assuming that the primary and secondary amines react completely with epoxide and that residual epoxide is consumed by epoxide-hydroxyl addition, all these reactions having approximately equal exothermicity. The calorimetric data of Gupta et al. can also be interpreted in this way.

Apicella et al.[100] obtained data on the same system for DDS concentrations in the range 0–100 parts per hundred parts by weight of resin (phr). The data from DSC scans was analysed by Prime's method[36] assuming overall first-order kinetics. The high temperature homopolymerisation of the resin occurred with a heat of reaction of 0.711 kJ/g resin (92.4 kJ/mole epoxide) and with an apparent activation energy of 171.5 kJ/mole, while the corresponding values for the mixture containing 100 phr of DDS were 1.067 kJ/g resin (138.7 kJ/mole) and 69.5 kJ/mole. A trend of increasing heat of reaction with increasing DDS concentration is reported. This was interpreted in terms of an increasing amount of the epoxide-primary addition relative to homopolymerisation, as the DDS concentration is increased. This conflicts with the observation mentioned above that the heat of reaction is approximately constant (108 to 110 kJ/mole) in the range 30–55 phr of DDS. Isothermal data in the range 140 to 205 °C for 35 phr of DDS were also obtained by Apicella et al. The initial rates at less than 10% conversion were used as an estimate of K_1 in Eq. (4-11).

Further isothermal data for this system is provided by Mijovic et al.[101]. Mixtures containing 23, 28, and 37 phr of DDS were examined in the temperature range 185 to 215 °C using hermetically sealed sample pans and assuming a horizontal baseline unter the DSC exotherms. After the completion of each isothermal experiment the samples were scanned at 10 K/min from 150 to 300 °C to determine any residual heat of reaction, and the sum of the isothermal and residual heat was taken as the overall heat of reaction, Q_0. The data were analysed in terms of the general autocatalytic kinetic equation, Eq. (4-11), assuming overall second-order kinetics, $m + n = 2$. At each temperature the rate constant K_1 was obtained from the initial rate, and then m and K_2 were obtained by the method of Ryan and Dutta[87] using the data at the maximum rate, and a reduced form of Eq. (4-11) for the maximum rate condition. For all three compositions the value of m was in the range 0.5–0.7, and although there is significant scatter in the data a positive correlation was observed between m and curing temperature. The values of Q_0 are given in units of J/g of reaction mixture and show a decrease with increasing DDS concentration, from 718 J/g at 23 phr of DDS to 581 J/g at 37 phr. However if an epoxide equivalent mass of 130 is assumed for the resin, the corresponding variation in Q_0 is 114.8 to 103.5 J/mole epoxide, in fair agreement with the constant value of about 110 kJ/mole for 30–55 phr of DDS[89]. The apparent activation energies associated with K_1 and K_2 were 91 and 133 kJ/mole for 23 phr of DDS, while the corresponding values for 37 phr were 66 and 50 kJ/mole. For the system containing 23 phr of DDS a small secondary exotherm peak was observed at about 40% conversion but it was not possible to deconvolute the overlapping data. This was thought to be qualitatively consistent with the observation of Morgan et al.[102], based on a spectroscopic (FTIR) study of a similar resin-DDS composition, that the rate of reaction of epoxide with primary amine becomes insignificant above 40% conversion, whereas the secondary amine and hydroxyl addition reactions with epoxide accelerate and continue to proceed to similar rates.

A kinetic model which includes both amine-epoxide and hydroxyl-epoxide addition reactions, with hydroxyl autocatalysis has been proposed by Zukas[103, 104]. The starting point was an expression for the rate of consumption of epoxide by reaction with primary or secondary amine and hydroxyl groups

$$-de/dt = k'_1 a_1 e + k'_2 a_2 e + k'_3 he + k_1 a_1 he + k_2 a_2 he + k_3 h^2 e \quad (4\text{-}14)$$

where e, a_1, a_2 and h are the concentrations of epoxide, primary amine, secondary amine, and hydroxyl, respectively. The first three terms on the right-hand side of Eq. (4-13) represent uncatalysed reactions and the remaining terms represent hydroxyl catalysed reactions. The relative rates of primary and secondary amine addition are $k_1/k_2 = k'_1/k'_2 = \beta$, and the corresponding ratio for primary amine relative to hydroxyl addition is $k_1/k_3 = k'_1/k'_3 = \lambda$.

An expression for overall rate in terms of fractional conversion was derived from Eq. (4-14), for an initial stoichiometric mixture of epoxide and amine. This was reduced to an applicable form by assuming that the concentration of ether groups from epoxide-hydroxyl addition was negligible, to give the Eq.:

$$d\alpha/dt = (K_1 + K_2\alpha)(1 - \alpha) f' \qquad (4-15)$$

where $\quad f' = 1 - \alpha_1 + (\alpha_1 - \alpha)/\beta + \alpha/\lambda \qquad (4-16)$

and $\alpha_1 = 1 - 2a_1/e_0$, e_0 being the initial epoxide concentration. This was extended to a more general form:

$$d\alpha/dt = (K_1 + K_2\alpha^m)(1 - \alpha)^n f' \qquad (4-17)$$

where m, n, β, and λ are used as adjustable empirical parameters for fitting experimental data to the Eq.

Zukas et al. [104)] studied the cure of TGDDM and diglycidylaniline (DGA) resins with stoichiometric proportions of either 4,4'-DDS or 3,3'-DDS. The reactions were monitored by isothermal DSC in the temperature range 140–210 °C using samples in hermetically sealed pans. The average total heat of cure was 104 ± 6 kJ/mole with no significant differences between resins or curing agents. For TGDDM and either

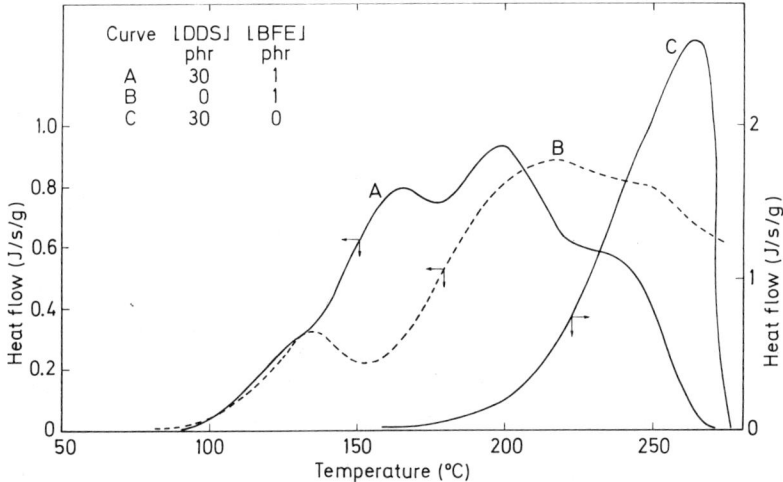

Fig. 12. DSC scans at 10 K/min heating rate showing the effect of BFE accelerator on the cure of TGDDM and TGDDM/DDS (© British Crown Copyright)

amine the apparent value of the reactivity ratio for secondary to primary amine with epoxide, $1/\beta$, was about 0,6, and for DGA the ratio was about 0.4. Values of λ obtained for the DGA reactions indicated that primary amine addition occurred at 20 to 80 times the rate of hydroxyl addition. The activation energy for the autoaccelerated reactions was in the range 62–65 kJ/mole.

In practice the epoxide-amine cure is often accelerated by the addition of catalysts such as boron trifluoride complexes, and the boron trifluoride-ethylamine adduct (BFE) is widely used for this purpose. In addition to catalysing the epoxide-amine reactions, BFE can initiate homopolymerisation of epoxide. The accelerating effect of BFE is illustrated by DSC scans for the TGDDM/DDS/BFE system in Figure 12. The multiple-peaked exotherm associated with the BFE-catalysed TGDDM/DDS cure indicates that the kinetics of this system are more complex than those for the cure with amine alone. For this system the overall heat of reaction was found to decrease with increasing BFE concentration [89]. For DDS alone Q_0 was about 110 kJ per mole epoxide while the value for BFE alone was 75 kJ/mole, and the DDS/BFE values were between these limits. It appears that the proportion of epoxide homopolymerisation relative to amine or hydroxyl addition increases with increasing BFE concentration.

Lee et al. [105] have reported a study of a commercial resin system, Hercules 3501-6, which is said to be a mixture of epoxy resins, of which the main component is TGDDM, with DDS curing agent and BFE accelerator in the weight proportions 100:34:1.5. The cure was monitored by DSC in both the temperature scanning and isothermal modes. DSC scans showed a double-peaked exotherm with a total area of 474 J/g. The isothermal data between 130 and 202 °C were fitted to Eq. (4-12) with B treated as an empirical constant (0.47 ± 0.07), at conversions up to 30%, using a non-linear regression method Above 30% conversion the data were found to fit simple first-order kinetics.

The activation energies related to K_1' and K_2' in Eq. (4-12) were 81 and 78 kJ/mole, respectively, while the first-order reaction at higher conversions had an activation energy of 57 kJ/mole.

Walkup et al. [114] have studied a BFE-catalysed system having the same composition as that reported for a commercial composite prepreg resin. The resin consisted of TGDDM (Ciba-Geigy MY 720), 64 wt.-%, diglycidylorthophthalate (Celanese Gly-Cel-A-100), 11%, DDS, 25%, and BFE, 0.4%. DSC scans were run on the individual components and systematic two, three and four-constituent mixtures in the same relative proportions as in the standard mixture. In addition the reactions of the individual constitutents and BFE were studied by proton, fluorine, and boron NMR spectroscopy. The DSC scan at 10 K/min for the complete resin system showed four overlapping peaks designated α, β, γ and δ. The α and β peaks had well-defined maxima at 240 and 200 °C, respectively, while the γ and δ peaks were defined as shoulders at 160 and 125 °C respectively. From the results of the DSC and NMR measurements it was proposed that the α peak is due to non-catalysed cure reactions, while the other peaks are associated with reactions involving the BFE catalyst. The α peak area decreased with increasing BFE concentration and approached zero at about 2 wt.-% BFE. The β peak showed a maximum area at about 0.4% BFE and decreased with higher concentrations of BFE, while the γ peak area increased monotonically with BFE concentration. The small δ peak was independent of BFE concentration and was even present in the absence of BFE. It was concluded that BFE can form a number of catalytic species the most stable of which is $BF_4^- \cdot NH_3^+ \cdot C_2H_5$, and that the β peak is associated with cationic catalysis of the cure by this species. Furthermore it was found that BFE reacts with epoxides to form monofluoroborates, and the γ-peak was attributed to monofluoroborate and BFE catalysis of cure reactions.

It was noted previously that a common feature of epoxy resin cure is the retardation of rate, observed particularly above the gel point and as the resin T_g approaches the cure temperature. It is likely that the reactions become controlled by the limited diffusion rates at high viscosity and high levels of chain entanglement and crosslinking. Gordon and Simpson [106] proposed that resin cross-linking reactions at high levels of conversion are controlled mainly by physical factors and that the rate would be described by the WLF equation, which can relate segmental motion to the difference between cure temperature and T_g. Lunak et al. [91] found for the system BADGE/4,4'-diamino-3,3'-dimethyldicyclohexylmethane, that the critical conversion at the gel point is independent of cure temperature and therefore also of the extent of diffusion control. They concluded that the decrease in reaction rate due to increasing viscosity affects all of the reaction steps to the same degree.

For the system BADGE/DDS in stoichiometric proportions, isothermal DSC cure rate data were obtained in the range 142–200 °C [107]. Plots were made of reduced rate, $(d\alpha/dt)/(1-\alpha)^n$, against α, in order to test the fit to Eq. (4-11) with m = 1, n = 1 or 2. In both cases the plots were approximately linear only at low conversion. For the case where n = 1, the plots consistently show an increasing retardation in reduced rate with increasing conversion, above 20–40% conversion. It was postulated that the retardation was associated with the increasing viscosity of the system and as a first approximation it was assumed that the apparent rate constants, K_1 and K_2 decrease linearly with increasing conversion, giving the modified Eq.

$$r = (K_1^* + K_2^*\alpha)(1-\alpha) \qquad (4-18)$$

The retarded rate parameters are given by

$$K_i^* = K_i - s_i\alpha \qquad (4-19)$$

for i = 1 or 2, and the constants s_i are

$$s_i = -dK_i^*/d\alpha \qquad (4-20)$$

Eq. (4-14) can then be written as

$$r = (A_0 + A_1\alpha + A_2\alpha^2)(1-\alpha) \qquad (4-21)$$

where $A_0 = K_1$, $A_1 = K_2 - s_1$ and $A_2 = -s_2$, so that a quadratic dependence of reduced rate on conversion is predicted.

The data were found to give a reasonably good fit to Eq. (4-21). The apparent rate constants K_1 and K_2 gave linear Arrhenius plots with apparent activation energies of 85 and 43 kJ/mole, respectively. A more detailed study of the inter-relationships between the chemical kinetics, the viscosity and the conversion could provide a useful insight into the nature of these diffusion-controlled reactions.

Table 3. Summary of DSC kinetic results on amine cure

Resin system [Epox]₀ [Amine]₀	Isothermal (I) or Dynamic (D) DSC	Kinetic Equation	E kJ/mole	ln A s	Ref.
BADGE/MPD[a] (DER332) 1.0/1.3 equiv.	D I (115–160 C)	(2-9) n = 0.9–1.3	50–63 47	8–12 8.8	73, 76
BADGE/DDM[b] (DER332) NARMCO 5208	D I (84–179 C) D	(2-6) (2-6) n = 1.7 +/− 0.2	53–59 50 109	— — —	77) 79)
PGE/BA[c] (various ratios)	I (50–70 C)	(4-10)	1) 58 2) 56	1) 11.6 2) 10.4	83)
BADGE/EDA[d] (DER 332LC) 0.9/1.0 equiv.	I (50–70 C)	(4-10)	1) 55	1) 10.8	83)
BADGE/TMDA[e] (1.0/1.0 equiv.)	I (50–70 C)	(4-10)	1) 57	1) 12.2	83)
BADGE/HMDA[f] 0.9/1.0 equiv.	I (50–70 C)	(4-10)	1) 54	1) 11.0	83)
BADGE/MPD (DER 332)	I (70–140 C)	(4-11) m = n = 1	1) 62 2) 48	1) 10.6 2) 9.1	85)
BADGE/MPD (DER 332) 1.0/1.0 equiv.	I (60–130 C)	(4-12) B = 1.0	1) 81 2) 48	1) 16.0 2) 9.5	86)
1.0/1.5 equiv.	I (60–130 C)	(4-12) B = 1.5	1) 88 2) 46	1) 18.9 2) 8.7	86)
BADGE/MPD (DER 332) 1.0/1.0 equiv.	I (90–170)	(4-11) m + n = 2	1) 65 2) 46	1) 11.8	87)
BADGE/MPD (Epon 826) 1.0/1.0 equiv.	I (100–140)	(4-11) m + n = 2	1) 64 2) 45	1) 11.7 2) 8.0	116)
BADGE/DDM (Epon 826) 1.0/1.0 equiv.	I (100–140)	(4-11) m + n = 2	1) 61 2) 44	1) 10.4 2) 7.7	116)
BADGE/BZA[g] (Epilox M515) 1.0/1.0 mol.	I (70–99)	(4-11) m ~ 1 n ~ 1.5	1) 56 2) 43	1) 11.0 2) 8.8	88)
BADGE/DBZEDA[h] 1.0/1.0 mol.	I (75–106)	(4-11) m ~ 1 n ~ 1.5	1) 62 2) 50	1) 12.7 2) 10.3	88)
BADGE/DDS[i] (Epikote 828) 1.0/1.0 equiv.	I (140–200)	(4-11) m = n = 1	1) 85 2) 43	1) 15.2 2) 5.3	107)
TGDDM/DDS (MY720)	I (170–220)	(4-11) m = n = 1	1) 80 2) 71	1) 13.4 2) 12.6	89)
TGDDM/DDS (Pure: >96%) 1.0/1.0 equiv.	I (140–200)	(4-16)	1) 76 2) 63		103, 104
TGDDM/3-DDS[j] 1.0/1.0 equiv.	I (140–200)	(4-16)	1) 75 2) 62		103, 104

Table 3. (continued)

Resin system [Epox]$_0$ [Amine]$_0$	Isothermal (I) or Dynamic (D) DSC	Kinetic Equation	E kJ/mole	ln A s	Ref.
DGA/DDS 1.0/1.0 equiv.	I (150–210)	(4-16)	1) 76 2) 65		103, 104)
DGA/3-DDS 1.0/1.0 equiv.	I (140–200)	(4-16)	1) 72 2) 60		103, 104)
TGDDM/DDS (MY720)					
100/0 pbw	D	(2-9)	172		100)
100/100 pbw	D	(2-9)	70		100)
100/35 pbw	I (140–205)	(4-11)	1) 73	1) 10.6^k	100)
TGDDM/DDS (MY720)					
100/23 pbw	I (185–215)	(4-11) m + n = 2 m = 0.5–0.7	1) 91 2) 133	1) 15.1^l 2) 25.8^l	101)
100/37 pbw	I (185–215)	(4-11)	1) 66 2) 50	1) 9.9^l 2) 6.9^l	101)

BADGE = Bisphenol-A-diglycidylether
PGE = Phenylglycidylether
TGDDM = Tetra-N-glycidyldiaminodiphenylmethane
DGA = Di-N-glycidylaniline

[a] Metaphenylenediamine
[b] 4,4'-Diaminodiphenylmethane
[c] Butylamine
[d] Ethylenediamine
[e] Trimethylenediamine
[f] Hexamethylenediamine
[g] Benzylamine
[h] N,N'-Dibenzylethylenediamine
[i] 4,4'-Diaminodiphenylsulphone
[j] 3,3'-Diamonodiphenylsulphone
[k] K from initial rate, E and A from Arrhenius plot in paper
[l] Calculated assuming original data in min

The results of DSC kinetic studies on systems utilising amine curing agents are summarised in Table 3. Again the general pattern is one of complexity in the kinetics. Autocatalysis and diffusion control are usually observed. Although simple kinetic models, represented by Eq. (2-6) to (2-9), may fit experimental data over a limited range of conversion, the kinetic parameters obtained from their use should only be regarded as empirical values applying under restricted conditions. Any extrapolation of such parameters to different ranges of temperature, concentration and conversion should be treated with caution.

4.3 Dicyanodiamide and Imidazoles as Curing Agents

4.3.1 Dicyanodiamide

The system comprising BADGE resin and dicyanodiamide (DICY) curing agent was studied by Sacher [60]. Two different resins were used, DER 332, and Epon 1001, the latter being a higher molecular weight form of BADGE. Mixtures of the resin and DICY were used with the equivalent ratio 1.1/1.0. Isothermal DSC data were obtained in the temperature range 170–220 °C; and fitted to Eq. (2-9). It was found that the rate depended on the particle size of the DICY. For finely ground DICY with particle size below 125 μm the apparent activation energy, E, was 101 kJ/mole, and values of reaction order, n, were in the range 0.7–1.4, whereas DICY with particle size 250–500 μm gave values of 51 kJ/mole for E and 1.0–2.1 for n. Evidence from IR spectra is given which indicates that melamine is produced during cure, from the decomposition of DICY. This appeared as an insoluble precipitate in the lower molecular weight resin (DER 332).

A series of epoxy resin formulations containing DICY as the curing agent with an accelerator denoted as TMBDA, has been examined by Abolafia [72], using dynamic DSC to monitor the cure. The samples were in the form of glass cloth preimpregnated with resin. The data were analysed in terms of Eqs. (2-9) and (2-15). Plots of log apparent rate constant against 1/T for assumed values of n in the range 1.0–2.8 were not linear above 180 °C and only approximately linear between 130 and 180 °C, which is an indication of the inadequacy of the kinetic model. For one formulation kinetic parameters obtained from DSC scans at 5, 10 and 20 K/min, related to the best approximation to Eq. (2-9) at lower temperatures are reported. The apparent activation energy was 117–120 kJ/mole but n and A appeared to depend on heating rate.

The cure kinetics of some epoxy resin powder coating composition were reported by Olcese et al. [108]. These were mixtures of BADGE resins with DICY and an epoxide-amine adduct or an imidazole as accelerator, together with TiO_2 and plasticisers. Data from DSC scans were analysed using Eq. (2-12) to obtain the apparent activation energy, E. Also Eq. (2-13) and (2-13a) were used to obtain estimates of E and order

Table 4. Kinetic analysis of DSC scan data for epoxy powder coating compositions (Ref. [108])

Composition[a]	E kJ/mole			n	
	Eq. (2-12)[b]	Eq. (2-13a)	Eq. (2-13)	Eq. (2-13a)	Eq. (2-13)
B	30–100	110	187	0.4	0.4
C	60–100	88	81	2.1	0.9
D	68– 95	79	81	2.2	1.0

[a] Resin BADGE (X102, Epoxide equivalent mass = 830), approx 62 wt.-%
  Curing agent: B, DICY/epoxy-amine adduct, 3.6 wt.-%
                C, DICY/imidazole derivative, 5.4 wt.-%
                D, as C, 2.9 wt.-%
[b] E decreases with increasing conversion

of reaction, n. Although the resin systems are complex formulations, the results do provide an interesting comparison of the different methods used for data analysis. Typical results are summarised in Table 4.

The variations in kinetic parameters obtained by the different methods are, in some cases, quite large, and it appears that the assumption of the simple kinetic function of Eq. (2-9), upon which Eqs. (2-13) and (2-13a) are based, is not justified for this system.

The cure of BADGE resin (DER 332) with DICY in the presence of 3-(p-chlorophenyl)-1,1-dimethylurea (Monuron) has been reported by Son and Weber [62]. A DSC scan at 10 K/min heating rate on a mixture of the resin and DICY in the weight ratio 100/10 showed that reaction started at 170 °C and the exotherm peak occurred at 200 °C. The three-component mixture, BADGE/DICY/Monuron (100/10/3) gave a large exotherm peak starting at 135 °C. This lower onset temperature indicates the accelerating effect of Monuron. A mixture of BADGE and Monuron (100/3) gave only a slight exotherm above 150 °C. This conflicts with the observation of LaLiberte et al. [63] who found that Monuron undergoes considerable reaction with BADGE at 90 °C. These workers also found that the DSC exotherm for such a mixture in an open sample pan, decreased in area with increasing heating rate, although scans at 0.5 K/min gave an apparent heat of reaction in agreement with isothermal measurement. It was proposed that this effect of heating rate and the observations of Son and Weber are explained by the evolution of dimethylamine from the sample. The formation of dimethylamine, which is an effective curing agent, from reaction involving Monuron has been observed by Son and Weber and by LaLiberte et al. and discussed in Sect. 3.2.

Further studies [63] were made on a resin blend containing predominantly an epoxy cresol novolak resin and BADGE, with DICY and Monuron in the wt.-% proportions, Resin/DICY/Monuron: 89/7.5/0, 89/7.5/3.8, and 89/0/3.8. Plots of log heating rate against reciprocal absolute temperature of the exotherm peak, from DSC scans, were found to be linear for all three compositions. The apparent activation energies estimated from the slopes of the plots were 161 kJ/mole, for DICY as the sole curing agent, 97 kJ/mole for the DICY/Monuron combination, and 80 kJ/mole for Monuron alone. It thus appears that the Monuron-epoxide reaction is significant in lowering the cure temperature of resin containing DICY.

Two resin systems utilising accelerated DICY cure have been studied by Schneider et al. [61]. The systems were BADGE/Epoxy cresol novolak /p-t-butylphenylglycidyl ether/DICY/Monuron in the wt.-% proportion 30/57/1.5/8/3.5, Resin 1, and TGDDM/epoxy cresol novolak/BADGE/DICY/Diuron, 70/14/5/3.5/3, Resin 2. Typical DSC scans at 5 K/min heating rate for the two resins are shown in Figs. 13 and 14.

The effective baselines shown in the Figs. 13 and 14 were obtained from second scans on the cured resins. In both cases the small endotherm at 206 °C was attributed to the melting of unreacted DICY. Three major exotherm peaks are seen in the DSC scan on Resin 2 (Fig. 14), in contrast to the single major peak for Resin 1 in Fig. 13, which indicates a more complex curing mechanism for the former. The peak at highest temperature for Resin 2 was associated with a rapid weight loss attributed to degradation reactions. The first exothermic peak in the DSC scan of uncured Resin 2 was absent in a scan of a sample previously cured at 170 °C for 16 minutes. A rough

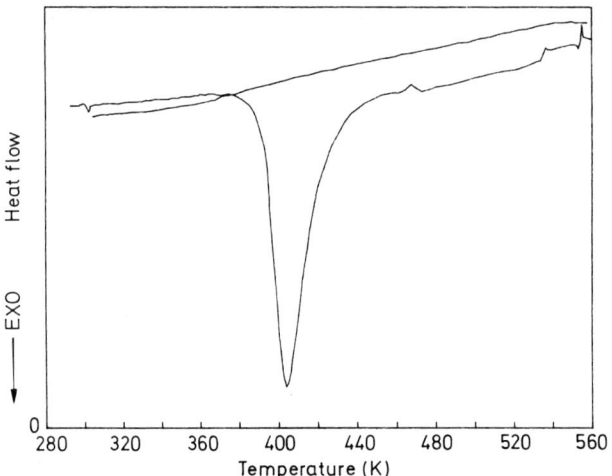

Fig. 13. DSC scans of Resin 1 at 5 K/min heating rate (From Ref. [61], Fig. 2)

resolution of the two peaks showed that about 38% additional cure occurs above 170 °C. Arrhenius plots for reciprocal time to 30% conversion, obtained from isothermal DSC in the temperature range 100–160 °C gave an apparent activation energy of 85 kJ/mole for both resin systems. This was in good agreement with values obtained from gelation times determined by Torsional Braid Analysis. In Resin 1 the amine/epoxide equivalent ratio is about 1.0 whereas it is about 0.25 for Resin 2. It was postulated that Resin 1 cures primarily by amine-epoxide additon reactions whereas in Resin 2 hydroxyl-epoxide addition makes a significant contribution to the cure, and the two reactions appear to have similar activation energies.

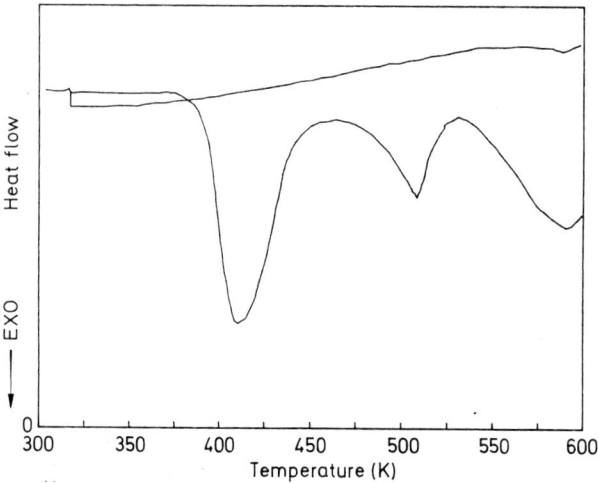

Fig. 14. DSC scans of Resin 2 at 5 K/min heating rate (From Ref. [61], Fig. 4)

Miller and Oebser [109] examined a series of BADGE/DICY compositions incorporating vaious hydroxylic or basic accelerators. The time to completion of cure was obtained at various temperatures from isothermal DSC. Apparent activation energies were obtained from linear Arrhenius plots using log cure time.

A commercial epoxy resin formulation used for resin-fibre composites, CIBA-GEIGY BSL 913, was investigated [110] using isothermal DSC in the range 90–143 °C. This system utilises an accelerated DICY cure. A master curve of conversion against log time was obtained by horizontal transposition of similar curves at other temperatures along the log time axis. The log time shift factors gave a linear Arrhenius plot from which an apparent activation energy of 90 kJ/mole was derived. A similar value, 87 kJ/mole, was obtained from gel time measurements.

The same author [111] studied a series of compositions consisting of TGDDM resin with micronised DICY and Diuron in various proportions, using DSC scans at four different heating rates. The scans obtained at 10 K/min heating rate are shown in Fig. 15, for the compositions containing TGDDM/DICY/Diuron in the weight ratios 100/10/0, 100/0/10, and 100/10/10. The samples were in open pans, and baselines were obtained by extrapolation of the linear region before exotherm onset. Repeat scans on cured samples showed that this was a reasonable approximation. The synergistic effect of Diuron is clearly seen in the shift of the main cure exotherm to lower temperatures. Although the DSC scans exhibit multiple peaks, which is symptomatic of a complex reaction, plots of $\ln r_\alpha$ against $1/T_\alpha$ (see Eq. (2-12)) from scans at different heating rates are remarkably linear. Typical examples of these plots, for the 100/10/10 mixture, are shown in Fig. 16. From the plots values of apparent activation energy, E, were obtained at different degrees of conversion. For this com-

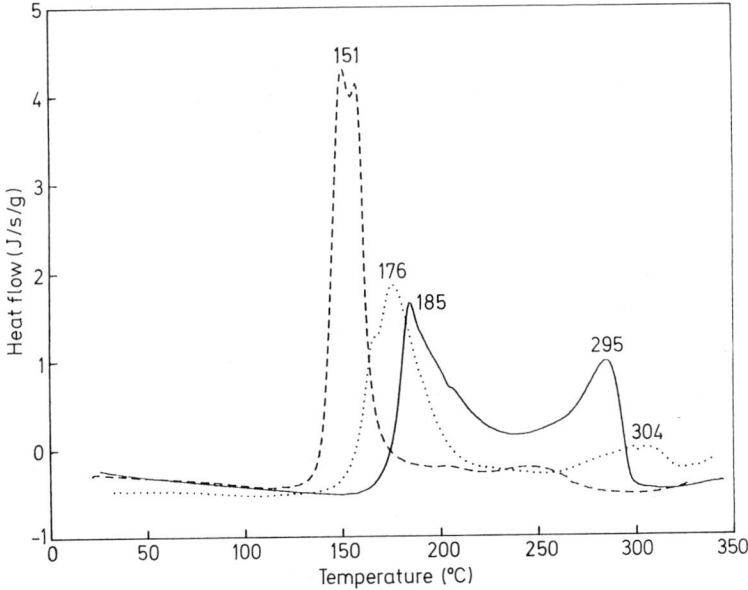

Fig. 15. DSC scans at 10 K/min heating rate for MY720/DICY/DIUR in the weight proportions: 100/10/0, ———; 100/0/10, – – – –; 100/10/10, – – – – (© British Crown Copyright)

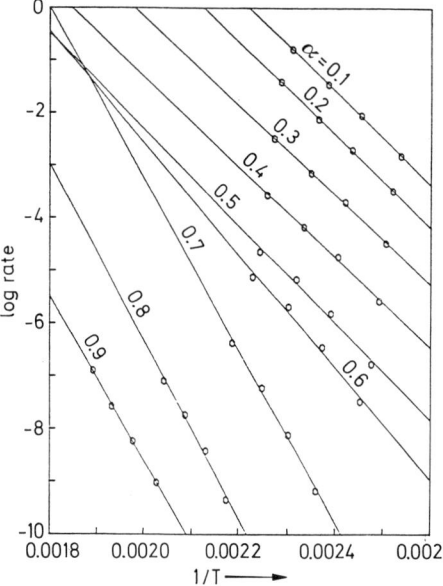

Fig. 16. Plots of ln r against 1/T for MY720/DICY/DIUR in the weight proportions 100/10/10. The plots for alpha = 0.1, 0.2, 0.3, 0.4 are displaces along the ordinate by 4, 3, 2, and 1 units respectively (© British Crown Copyright)

position E is 70 kJ/mole in the conversion range 10–50% and it increases to 128 kJ per mole at 70–90% conversion, and a similar trend was observed for the other compositions investigated. The addition of Diuron to the Resin/DICY mixture had the effect of lowering E at 30% conversion from 89 to 68 kJ/mole while the corresponding value for Resin/Diuron alone was 17 kJ/mole.

The results of the DSC studies on DICY-cured systems are summarised in Table 5. The general comments at the end of Section 4.2, on the complexity of the kinetics apply also to dicyanodiamide cure. In this case the systems are even more complex because the dicyanodiamide is initially present as a dispersion of crystalline particles, so that the onset of the curing reaction is usually coincident with the onset of dissolution of the dicyanodiamide.

4.3.2 Imidazoles

Another type of curing agent of technological importance is imidazole and its derivatives which can promote base-catalysed homopolymerisation of epoxide. However very few DSC studies have been reported on well-defined systems based on imidazoles.

The kinetics of formation of 1:1 and 2:1 molar adducts of 2-ethyl-4-methylimidazole (EMI) with PGE and BADGE resin have been investigated [37] using isothermal and dynamic DSC. Isothermal data for the reaction of PGE and EMI in the 2:1 molar ratio were obtained at 80, 99, and 120 °C. Heat flow data were derived assuming a linear baseline and analysed using the integrated form of Eq. (2-9). Plots of $\log(1 - \alpha)$ against time were linear, indicating a first-order reaction. Dynamic data obtained at a heating rate of 20 K/min were analysed in terms of Eq. (2-9). Linear Arrhenius plots for first-order kinetics were obtained from both the original data, and the data corrected for heating rate by Prime's method, Eq. (2-23). However the corrected data gave better agreement with the isothermal values for E and A. The activation

Table 5. Summary of DSC kinetic results on dicyanodiamide cure

Resin system	Isothermal (I) or Dynamic (D) DSC	Kinetic Equation	E kJ/mole	ln A s^{-1}	Ref.
BADGE/DICY (DER332) 1.0/0.9 equiv. DICY particle size (μm)	I (170–220 C)	(2-9)			60)
<125		n = 0.7–1.4	101	22.1	
250–500		n = 1.0–2.1	51	7.8	
BADGE + ECN/DICY/ MON (EPON 828)					63)
89/7.5/0 wt.-%	D (170–200)	(2-14)	161		
89/7.5/3.8 wt.-%	D (130–160)	(2-14)	97		
89/0/3.8 wt.-%	D (130–170)	(2-14)	80		
BADGE/ECN/BPGE/ DICY/MON (EPON 828)					
30/57/1.5/8/3.5 wt.-%	I (100–150)	a	85		61)
TGDDM/ECN/BADGE/ DICY/DIUR					
30/57/1.5/8/3.5 wt.-%	I (100–160)	a	85		61)
BADGE/DICY 1.0/2.0 equiv.	I (150–200)	b	69		109)
BADGE/DICY/MI 1.0/0.7–1.6/0.05 equiv.	I (150–200)	b	66–70		109)
BADGE/DICY/OH 1.0/0.5/0.16 equiv.	I (150–200)	b	51		109)
BADGE/DICY/CADP 1.0/0.6/0.04 equiv.	I (150–200)	b	35		109)
TGDDM/DICY/DIUR (MY720)					
100/5/0 pbw	D (190–240)	(2-6, 2-12)	84 (at 30% conversion)		111)
100/5/10	D (130–170)	(2-6, 2-12)	54 (at 30% conversion)		111)
100/10/0	D (175–220)	(2-6 2-12)	89 (at 30% conversion)		111)
100/10/10	D (125–170)	(2-6, 2-12)	68 (at 30% conversion)		111)
100/0/10	D (140–205)	(2-6, 2-12)	17 (at 30% conversion)		111)

DICY = Dicyanodiamide
ECN = Epoxy-cresol-novolak resin
MON = Monuron
BPGE = t-Butylphenylglycidyl ether
DIUR = Diuron
MI = 2-Methylimidazole
OH = Unspecified hydroxyl
CADP = Carboxylic acid salts of dimethyl-aminomethylphenol

^{a} Plot of ln (1/(time to 30% conversion)) vs. 1/T
^{b} Plot of log (time to cure) vs. 1/T

energy for the formation of PGE · (EMI)$_2$, from isothermal measurements was 82 kJ/mole and the first-order rate constant at 99 °C was 7.46×10^{-3} s^{-1}. A further isothermal DSC study on the PGE/EMI system [112], utilised the non-linear baseline, obtained from a repeat run on the reacted sample, to extend the measurements to higher temperatures, up to 150 °C. The use of the non-linear baseline gave an approximately constant heat of reaction, 460 J/g, with a standard deviation of 17 J/g, in the temperature range 80–150 °C. In contrast the assumption of a horizontal baseline indicated an apparent decrease in heat of reaction with increasing temperature. At 151 °C the first-order plot was linear above 50% conversion, giving a rate constant of 0.113 s^{-1}. The apparent induction period in the early part of the reaction may be due to complexity in the reaction mechanism at this temperature. The same reaction has also been studied by Grentzer et al. [113], using isothermal DSC in the range 80 to 120 °C, and dynamic DSC. The apparent activation energy obtained from single DSC scans using Eq. (2-9) was higher than the isothermal value. The isothermal heat of reaction decreased with increasing temperature, from 444 J/g at 80 °C to 290 J/g at 120 °C, and for this reason the authors consider the data to be unreliable. However the dynamic data from multiple scans analysed by the methods of Kissinger [23], Equation (2-14), and Ozawa [28-30], Eq. (2-22), yielded apparent activation energies significantly lower than the isothermal or single dynamic scan method. An evaluation of the three methods used for analysis of the dynamic data was made using the decomposition of 2,2'-azo-bis-isobutyronitrile dissolved in n-butyl phthalate. Results are available for the kinetics of this reaction, obtained by chemical analysis. It was found that the Kissinger and Ozawa methods gave values of E and ln A more than 25% lower than the literature data, while the results of the single dynamic scan method were within 5%. For this reason the authors considered that the single dynamic scan method would give the most consistent and reliable results for the EMI/PGE reaction. The modified isothermal method [112] using the true baseline may be the preferred technique.

Table 6. Summary of DSC kinetic results on imidazole-epoxide reactions

Resin system	Isothermal (I) or Dynamic (D) DSC	Kinetic Equation	E kJ/mole	ln A s	Ref.
PGE/EMI	I (80–220 C)	(2-9)	82	21.6	37)
1.0/2.0 mol	I (80–151 C)	n = 1	(k = 0.113 s at 151 C)		112)
	D	n = 1	88–93	21.4–23.0	37)
	I (80–120 C)	n = 1	84	22	113)
	D	n = 1	108–126	30–35	113)
	D	(2-14)	60	18	113)
	D	(2-22)	67	20	113)
PGE/EMI 1.0/1.0 mol	D	(2-22)	81	19.6	37)
BADGE/EMI (Epikote 825)					
1.0/1.0 mol	D	(2-22)	88	21.6	37)
1.0/2.0 mol	D	(2-22)	88	21.9	37)

EMI = 2,4-Ethylmethylimidazole

An unpublished study by the present author on the cure of BADGE resin with EMI (2–9 wt.-%) revealed multiple peaks in the DSC thermograms, indicative of a complex mechanism.

The DSC kinetic data for imidazole-epoxide reactions are summarised in Table 6.

5 Conclusions

The value of DSC as a method of monitoring exothermic epoxy resin curing reactions has been established. The main advantages of the technique are the relative ease of operation and the requirement for only small samples, of the order of 10 mg. As a means of following the overall course of the curing reaction through changes in heat evolution, or changes in glass transition temperature, it is very useful. However its use is not without potential problems. In particular if the reaction is complex in mechanism the deconvolution of the experimental data in terms of the underlying reactions can be very difficult. Also in such cases the validity of the basic assumptions underlying the application of DSC should be considered. Examples of the uncritical use of over-simplified kinetic models can be found in the literature.

The trend towards the analysis of DSC kinetic data by computer while offering advantages also brings its own hazards, especially if standard programs are applied to experimental data without an assessment of the relevance of the basic theoretical model to the reaction under investigation.

Although DSC scans are relatively easy and quick in execution it is advisable to also obtain isothermal data at several different temperatures. Although the extraction of kinetic data from a single DSC scan is feasible in principle, it can be misleading if the reaction is complex.

DSC is increasingly being applied to the study of epoxy resin cure in combination with other analytical methods such as nuclear magnetic resonance and Fourier transform infra-red spectroscopy, chromatographic methods, and dynamic mechanical or dielectric studies. It is probably as part of such combined investigations that DSC can be used most effectively in basic research, and in quality control and assessment.

6 References

1. Mackenzie, R. C.: Thermochim. Acta 28, 1 (1979)
2. Mackenzie, R. C.: Anal. Proc. 217 (1980)
3. Watson, E. S. O'Neill, M. J. Justin, J., Brenner, N.: Anal. Chem. 36, 1233 (1964)
4. O'Neill, M. J.: Anal. Chem. 36, 1238 (1964)
5. Boersma, S. L.: J. Amer. Ceramic Soc. 38, 281 (1955)
6. Lee, J. D., Levy, P. F.: North American Thermal Anal. Soc., Proc., 11th Conf. (1981)
7. Wendlandt, W. W., Gallagher, P. K.: Thermal Characterisation of Polymeric Materials (ed.) Turi, E. A., Chapter 1, New York N.Y., Academic Press (1981)
8. Mraw, S. C.: Rev. Sci. Instr. 53, 228 (1982)
9. Dallek, S., Kabacoff, L.: Thermochim. Acta 57, 99 (1982)
10. Hohne, G. W. H., Breuer, K. H., Eysel, W.: Thermochim. Acta 69, 145 (1983)
11. Richardson, M. J., Savill, N. G.: Polymer 16, 753 (1975)

12. Flynn, J. H.: Thermochim. Acta *8*, 69 (1974)
13. Nielsen, L. E.: J. Macromol. Sci., Revs. Macromol. Chem. *C3*, 69 (1969)
14. Nielsen, L. E.: Mechanical Properties of Polymers and Composites, Vol. 1, New York, N.Y., Marcel Dekker (1974)
15. Chompff, A. J.: Polymer Networks (ed.) Chompff, A. J., Newman, S., p. 145, New York, N.Y., Plenum Press (1971)
16. Widmann, G.: Proc. 4th International Conference on Thermal Analysis (ed.) Buzas, I., Vol. 3, p. 359, Budapest, Akademiai Kiado (1975)
17. Barton, J. M.: Thermochim. Acta *71*, 337 (1983)
18. Barrett, K. E. J.: J. Appl. Polym. Sci. *11*, 1617 (1967)
19. Freeman, E. S., Carroll, B.: J. Phys. Chem. *62*, 394 (1958)
20. Ellerstein, S. M.: Analytical Calorimetry (eds.) Porter, R. S., Johnson, J. O., p. 279, New York, N.Y., Plenum Press (1968)
21. Crane, L. W., Dynes, P. J., Kaelble, D. H.: J. Polym. Sci., Polym. Letters *11*, 533 (1973)
22. Rogers, R. N., Smith, L. C.: Thermochim. Acta *1*, 1 (1970)
23. Kissinger, H.: Anal. Chem. *21*, 1702 (1957)
24. Doyle, C. D.: J. Appl. Polym. Sci. *6*, 639 (1962)
25. Doyle, C. D.: Techniques and Methods of Polymer Evaluation (ed.) Slade, P. E., Jenkins, L. T., Vol. 1, Thermal Analysis Chapter *4*, p. 173, New York, N.Y., Marcel Dekker (1966)
26. Horowitz, H. H., Metzger, G.: Anal. Chem. *35*, 1464 (1963)
27. Coates, A. W., Redfern, J. P.: Nature *201*, 68 (1964)
28. Ozawa, T.: Bull. Chem. Soc. Japan *38*, 1881 (1965)
29. Ozawa, T.: J. Thermal Anal. *2*, 301 (1970)
30. Ozawa, T.: J. Thermal. Anal. *7*, 601 (1975)
31. Swarin, S. J., Wims, A. M.: Anal. Calorimetry *4*, 155 (1976)
32. Draper, A. L.: Third Toronto Symposium on Thermal Analysis, Proceedings p. 63 (1970)
33. MacCallum, J. R., Tanner, J.: Nature *225*, 1127 (1970)
34. Fevre, A. Murat, M., Comel, C.: J. Thermal Anal. *12*, 193 (1977)
35. Prime, R. B.: Analytical Calorimetry *2*, 201 (1970)
36. Prime, R. B.: Polym. Eng. Sci. *13*, 365 (1973)
37. Barton, J. M., Shepherd, P. M.: Makromol. Chem. *176*, 919 (1975)
38. Hill, R. A.: Nature *227*, 703 (1970)
39. Simmons, E. L., Wendlandt, W. W.: Thermochim. Acta *3*, 498 (1972)
40. Gyulai, G., Greenhow, E. J.: Thermochim. Acta: *5*, 481 (1973)
41. Sestak, J., Kratochvil, J.: J. Thermal Anal. *5*, 429 (1973)
42. Tanaka, Y., Mika, T. F.: Epoxide curing reactions, in: Epoxy Resins (ed.) May, C. A., Tanaka, Y., Chapter *3*, New York, N.Y., Marcel Dekker 1973
43. Mika, T. F.: Curing agents and modifiers, in: Epoxy Resins, Reference 42, Chapter 4
44. Fisch, W., Hofmann, W.: J. Polym. Sci. *12*, 497 (1954)
45. Fisch, W., Hofmann, W.: Makromol. Chem. *44–46*, 8 (1961)
46. Tanaka, Y., Kakiuchi, H.: J. Appl. Polym. Sci. *7*, 1063 (1963)
47. Tanaka, Y., Kakiuchi, H.: J. Polym. Sci., Part A, *2*, 3405 (1964)
48. Tanaka, Y., Kakiuchi, H.: J. Macromol. Chem. *1*, 307 (1966)
49. Luston, J., Manasek, Z.: J. Macromol. Sci., Chem. A, *12*, 983 (1978)
50. Luston, J., Manasek, Z., Kulichova, M.: J. Macromol. Sci., Chem. A, *12*, 995 (1978)
51. Matejka, L., Lovy, J., Pokorny, S., Bouchal, K., Dusek, K.: J. Polym. Sci., Polym. Chem. Ed. *21*, 2873 (1983)
52. Stevens, G. C.: J. Appl. Polym. Sci. *26*, 4259 (1981)
53. Stevens, G. C.: J. Appl. Polym. Sci. *26*, 4279 (1981)
54. Antoon, M. K., Koenig, J. L.: J. Polym. Sci., Polym. Chem. Ed. *19*, 549, (1981)
55. Schechter, L., Wynstra, J., Kurkjy, R. P.: Ind. Eng. Chem. *48*, 94 (1956)
56. Enikolopiyan, N. S.: Pure and Appl. Chem. *48*, 317 (1976)
57. Smith, I. T.: Polymer *2*, 95 (1961)
58. King, J. J., Bell, J. P.: in: Epoxy Resin Chemistry (ed.) Bauer R. S., ACS Symposium Series, No. 114, Chapter 16, Amer. Chem. Soc. (1979)
59. Saunders, T. F., Levy, M. F., Serefino, J. F.: J. Polym. Sci. A-1, *5*, 1609 (1967)
60. Sacher, E.: Polymer *14*, 91 (1973)

61. Schneider, N. S., Sprouse, J. F., Hagnauer, G. L., Gillham, J. K.: Polym. Eng. Sci. *19*, 304 (1979)
62. Son, P., Weber, C. D.: J. Appl. Polym. Sci. *17*, 1305 (1973)
63. LaLiberte, B. R., Bornstein, J., Sacher, R. E.: Ind. Eng. Chem. Prod. Res. Dev. *22*, 261 (1983)
64. Farkas, A., Strohm, P. F.: J. Appl. Polym. Sci. *12*, 159 (1968)
65. Fava, R. A.: Polymer *9*, 137 (1968)
66. Peyser, P., Bascom, W. D.: Analytical Calorimetry *3*, 537 (1974)
67. Peyser, P., Bascom, W. D.: J. Appl. Polym. Sci. *21*, 2359 (1977)
68. Mohler, H., Schwab, M.: Kunststoffe *71*, 245 (1981)
69. Malavasic, T., Moze, A., Vizovisek, I., Lapanje, S.: Angew. Makromol. Chemie *44*, 89 (1975)
70. Stevens, G. C., Richardson, M. J.: Polymer *24*, 851 (1983)
71. Duswalt, A. A.: Thermochim. Acta *8*, 57 (1974)
72. Abolafia, O. R.: SPE Ann. Tech. Conf., Proc. *15*, 610 (1969)
73. Prime, R. B.: Analytical Calorimetry *2*, 201 (1970)
74. Acitelli, M. A., Prime, R. B., Sacher, E.: Polymer *12*, 335 (1971)
75. Prime, R. B., Sacher, E.: Polymer *13*, 455 (1972)
76. Prime, R. B.: Polym. Eng. Sci. *13*, 365 (1973)
77. Barton, J. M.: Makromol. Chem. *171*, 247 (1973)
78. Barton, J. M.: J. Macromol. Sci.-Chem. *A8*, 25 (1974)
79. Cizmecioglu, M., Gupta, A.: SAMPE Qtly. April, 16 (1982)
80. Levy, N.: Analysis of an epoxy resin curing reaction by differential scanning calorimetry, in: Computer Applications in Applied Polymer Science, p. 313, ACS Symp. Ser. 197, Washington D.C., Amer. Chem. Soc. 1975
81. Cizmecioglu, M., Hong, S. D., Moacanin, J., Gupta, A.: ACS Polymer Preprints *22*, (2), 224 (1981)
82. Severini, F., Gallo, R., Marullo, R.: J. Thermal Anal. *25*, 515 (1982)
83. Horie, K., Hiura, H., Sawada, M., Mita, I., Kambe, H.: J. Polym. Sci., Part A-1 *8*, 1357 (1970)
84. Klute, C. H., Viehmann, W.: J. Appl. Polym. Sci. *5*, 86 (1961)
85. Kamal., M. R., Sourour, S., Ryan, M.: SPE 31st Ann. Tech. Conf., Proc. *187* (1973)
86. Sourour, S., Kamal., M. R.: Thermochim. Acta *14*, 41 (1976)
87. Ryan, M. E., Dutta, A.: Polymer *20*, 203 (1979)
88. Flammersheim, H., Horhold, H., Bellstedt, K., Klee, J.: Makromol. Chem. *184*, 113 (1983)
89. Barton, J. M.: Unpublished work
90. Lunak, S., Dusek, K.: J. Polym. Sci., Polym. Symp. No. 53, 45 (1975)
91. Lunak, S., Vladyka, J., Dusek, K.: Polymer *19*, 931 (1978)
92. Dobas, I., Eichler, J.: Coll. Czech. Chem. Commun. *38*, 2602 (1973)
93. Dobas, I., Eichler, J.: Coll. Czech. Chem. Commun. *38*, 3279 (1973)
94. Dobas, I., Eichler, J.. Klaban, J.: Coll. Czech. Chem. Commun. *40*, 2989 (1975)
95. Johncock, P., Tudgey, G. F.: Brit. Polym. J. *15*, 14 (1983)
96. Johncock, P., Porecha, L., Tudgey, G. F.: J. Polym. Sci., Polym. Chem. Ed. *23*, 291 (1985)
97. Mones, E. T., Morgan, R. J.: ACS Polymer Preprints *22*, 249 (1981)
98. Hagnauer, G. L., Pearce, P. J., LaLiberte, B. R., Roylance, M. E.: ACS Div. Org. Coat. Plast. Chem., Preprints *47*, 429 (1982)
99. Gupta, A., Cizmecioglu, D., Coulter, R., Liang, R. H., Yavrouian, A., Tsay, F. D., Moacanin, J.: J. Appl. Polym. Sci. *28*, 1011 (1983)
100. Apicella, A., Nicolais, L., Iannone, M., Passerini, P.: J. Appl. Polym. Sci. *29*, 2083 (1984)
101. Mijovic, J., Kim, J., Slaby, J.: J. Appl. Polym. Sci. *20*, 1449 (1984)
102. Morgan, R. J., Happe, J. A., Mones, E. T.: Paper presented at the 28th Nat. SAMPE Symposium, Annaheim, Calif., April (1983)
103. Zukas, W. X.: Ph. D Dissertation, Chem. Eng. Dept., Univ. of Mass., USA (1983)
104. Zukas, W. X., Schneider, N. S., MacKnight, W. J.: Paper presented at Fall Meeting, ACS, Washington DC (1983)
105. Lee, W. I., Loos, A. C., Springer, G. S.: J. Comp. Mat. *16*, 510 (1982)
106. Gordon, M., Simpson, W.: Polymer *2*, 303 (1961)
107. Barton, J. M.: Polymer *21*, 603 (1980)
108. Olcese, T., Spelta, O., Vargiu, S.: J. Polym. Sci., Polym. Symp. No. 53, 113 (1975)
109. Miller, R. L., Oebser, M. A.: Thermochim. Acta *36*, 121 (1980)
110. Barton, J. M.: Brit. Polym. J. *11*, 115 (1979)

111. Barton, J. M.: Unpublished work
112. Barton, J. M.: Thermochim. Acta 71, 337 (1983)
113. Grentzer, T. H., Holsworth, R. M., Provder, T., Kline, S.: ACS Polymer Preprints 22, (1), 318 (1981)
114. Walkup, C. M., Morgan, R. J., Hoheisel, T. H.: ACS Polymer Preprints 25, (1), 187 (1984)
115. Riccardi, C. C., Adabbö, H. E., Williams, R. J. J.: J. Appl. Polym. Sci. 29, 2841 (1984)
116. Foun, C. C., Moroni, A., Pearce, E. M., Mijovic, J.: Polym. Mater. Sci. Eng. 51, 411 (1984)

K. Dusek (Editor)
Received January 9, 1985

Author Index Volumes 1–72

Allegra, G. and *Bassi, I. W.:* Isomorphism in Synthetic Macromolecular Systems. Vol. 6, pp. 549–574.
Andrews, E. H.: Molecular Fracture in Polymers. Vol. 27, pp. 1–66.
Anufrieva, E. V. and *Gotlib, Yu. Ya.:* Investigation of Polymers in Solution by Polarized Luminescence. Vol. 40, pp. 1–68.
Apicella, A. and *Nicolais, L.:* Effect of Water on the Properties of Epoxy Matrix and Composite. Vol. 72, pp. 69–78.
Apicella, A., Nicolais, L. and *de Cataldis, C.:* Characterization of the Morphological Fine Structure of Commercial Thermosetting Resins Through Hygrothermal Experiments. Vol. 66, pp. 189–208.
Argon, A. S., Cohen, R. E., Gebizlioglu, O. S. and *Schwier, C.:* Crazing in Block Copolymers and Blends. Vol. 52/53, pp. 275–334
Arridge, R. C. and *Barham, P. J.:* Polymer Elasticity. Discrete and Continuum Models. Vol. 46, pp. 67–117.
Aseeva, R. M., Zaikov, G. E.: Flammability of Polymeric Materials. Vol. 70, pp. 171–230.
Ayrey, G.: The Use of Isotopes in Polymer Analysis. Vol. 6, pp. 128–148.

Bässler, H.: Photopolymerization of Diacetylenes. Vol. 63, pp. 1–48.
Baldwin, R. L.: Sedimentation of High Polymers. Vol. 1, pp. 451–511.
Balta-Calleja, F. J.: Microhardness Relating to Crystalline Polymers. Vol. 66, pp. 117–148.
Barton, J. M.: The Application of Differential Scanning Calorimetry (DSC) to the Study of Epoxy Resins Curing Reactions. Vol. 72, pp. 111–154.
Basedow, A. M. and *Ebert, K.:* Ultrasonic Degradation of Polymers in Solution. Vol. 22, pp. 83–148.
Batz, H.-G.: Polymeric Drugs. Vol. 23, pp. 25–53.
Bekturov, E. A. and *Bimendina, L. A.:* Interpolymer Complexes. Vol. 41, pp. 99–147.
Bergsma, F. and *Kruissink, Ch. A.:* Ion-Exchange Membranes. Vol. 2, pp. 307–362.
Berlin, Al. Al., Volfson, S. A., and *Enikolopian, N. S.:* Kinetics of Polymerization Processes. Vol. 38, pp. 89–140.
Berry, G. C. and *Fox, T. G.:* The Viscosity of Polymers and Their Concentrated Solutions. Vol. 5, pp. 261–357.
Bevington, J. C.: Isotopic Methods in Polymer Chemistry. Vol. 2, pp. 1–17.
Bhuiyan, A. L.: Some Problems Encountered with Degradation Mechanisms of Addition Polymers. Vol. 47, pp. 1–65.
Bird, R. B., Warner, Jr., H. R., and *Evans, D. C.:* Kinetik Theory and Rheology of Dumbbell Suspensions with Brownian Motion. Vol. 8, pp. 1–90.
Biswas, M. and *Maity, C.:* Molecular Sieves as Polymerization Catalysts. Vol. 31, pp. 47–88.
Biswas, M., Packirisamy, S.: Synthetic Ion-Exchange Resins. Vol. 70, pp. 71–118.
Block, H.: The Nature and Application of Electrical Phenomena in Polymers. Vol. 33, pp. 93–167.
Bodor, G.: X-ray Line Shape Analysis. A Means for the Characterization of Crystalline Polymers. Vol. 67, pp. 165–194.
Böhm, L. L., Chmeliř, M., Löhr, G., Schmitt, B. J. and *Schulz, G. V.:* Zustände und Reaktionen des Carbanions bei der anionischen Polymerisation des Styrols. Vol. 9, pp. 1–45.

Bovey, F. A. and *Tiers, G. V. D.:* The High Resolution Nuclear Magnetic Resonance Spectroscopy of Polymers. Vol. 3, pp. 139–195.

Braun, J.-M. and *Guillet, J. E.:* Study of Polymers by Inverse Gas Chromatography. Vol. 21, pp. 107–145.

Breitenbach, J. W., Olaj, O. F. und *Sommer, F.:* Polymerisationsanregung durch Elektrolyse. Vol. 9, pp. 47–227.

Bresler, S. E. and *Kazbekov, E. N.:* Macroradical Reactivity Studied by Electron Spin Resonance. Vol. 3, pp. 688–711.

Bucknall, C. B.: Fracture and Failure of Multiphase Polymers and Polymer Composites. Vol. 27, pp. 121–148.

Burchard, W.: Static and Dynamic Light Scattering from Branched Polymers and Biopolymers. Vol. 48, pp. 1–124.

Bywater, S.: Polymerization Initiated by Lithium and Its Compounds. Vol. 4, pp. 66–110.

Bywater, S.: Preparation and Properties of Star-branched Polymers. Vol. 30, pp. 89–116.

Candau, S., Bastide, J. and *Delsanti, M.:* Structural. Elastic and Dynamic Properties of Swollen Polymer Networks. Vol. 44, pp. 27–72.

Carrick, W. L.: The Mechanism of Olefin Polymerization by Ziegler-Natta Catalysts. Vol. 12, pp. 65–86.

Casale, A. and *Porter, R. S.:* Mechanical Synthesis of Block and Graft Copolymers. Vol. 17, pp. 1–71.

Cerf, R.: La dynamique des solutions de macromolecules dans un champ de vitesses. Vol. 1, pp. 382–450.

Cesca, S., Priola, A. and *Bruzzone, M.:* Synthesis and Modification of Polymers Containing a System of Conjugated Double Bonds. Vol. 32, pp. 1–67.

Chiellini, E., Solaro R., Galli, G. and *Ledwith, A.:* Pptically Active Synthetic Polymers Containing Pendant Carbazolyl Groups. Vol. 62, pp. 143–170.

Cicchetti, O.: Mechanisms of Oxidative Photodegradation and of UV Stabilization of Polyolefins. Vol. 7, pp. 70–112.

Clark, D. T.: ESCA Applied to Polymers. Vol. 24, pp. 125–188.

Coleman, Jr., L. E. and *Meinhardt, N. A.:* Polymerization Reactions of Vinyl Ketones. Vol. 1, pp. 159–179.

Comper, W. D. and *Preston, B. N.:* Rapid Polymer Transport in Concentrated Solutions. Vol. 55, pp. 105–152.

Corner, T.: Free Radical Polymerization — The Synthesis of Graft Copolymers. Vol. 62, pp. 95–142.

Crescenzi, V.: Some Recent Studies of Polyelectrolyte Solutions. Vol. 5, pp. 358–386.

Crivello, J. V.: Cationic Polymerization — Iodonium and Sulfonium Salt Photoinitiators, Vol. 62, pp. 1–48.

Davydov, B. E. and *Krentsel, B. A.:* Progress in the Chemistry of Polyconjugated Systems. Vol. 25, pp. 1–46.

Dettenmaier, M.: Intrinsic Crazes in Polycarbonate Phenomenology and Molecular Interpretation of a New Phenomenon. Vol. 52/53, pp. 57–104

Dobb, M. G. and *McIntyre, J. E.:* Properties and Applications of Liquid-Crystalline Main-Chain Polymers. Vol. 60/61, pp. 61–98.

Döll, W.: Optical Interference Measurements and Fracture Mechanics Analysis of Crack Tip Craze Zones. Vol. 52/53, pp. 105–168

Dole, M.: Calorimetric Studies of States and Transitions in Solid High Polymers. Vol. 2, pp. 221–274.

Dorn, K., Hupfer, B., and *Ringsdorf, H.:* Polymeric Monolayers and Liposomes as Models for Biomembranes How to Bridge the Gap Between Polymer Science and Membrane Biology? Vol. 64, pp. 1–54.

Dreyfuss, P. and *Dreyfuss, M. P.:* Polytetrahydrofuran. Vol. 4, pp. 528–590.

Drobnik, J. and *Rypáček, F.:* Soluble Synthetic Polymers in Biological Systems. Vol. 57, pp. 1–50.

Dröscher, M.: Solid State Extrusion of Semicrystalline Copolymers. Vol. 47, pp. 120–138.

Dušek, K. and *Prins, W.:* Structure and Elasticity of Non-Crystalline Polymer Networks. Vol. 6, pp. 1–102.
Duncan, R. and *Kopeček, J.:* Soluble Synthetic Polymers as Potential Drug Carriers. Vol. 57, pp. 51–101.

Eastham, A. M.: Some Aspects of the Polymerization of Cyclic Ethers. Vol. 2, pp. 18–50.
Ehrlich, P. and *Mortimer, G. A.:* Fundamentals of the Free-Radical Polymerization of Ethylene. Vol. 7, pp. 386–448.
Eisenberg, A.: Ionic Forces in Polymers. Vol. 5, pp. 59–112.
Eiss, N. S. Jr. see Yorkgitis, E. M. Vol. 72, pp. 79–110.
Elias, H.-G., Bareiss, R. und *Watterson, J. G.:* Mittelwerte des Molekulargewichts und anderer Eigenschaften. Vol. 11, pp. 111–204.
Elsner, G., Riekel, Ch. and *Zachmann, H. G.:* Synchrotron Radiation Physics. Vol. 67, pp. 1–58.
Elyashevich, G. K.: Thermodynamics and Kinetics of Orientational Crystallization of Flexible-Chain Polymers. Vol. 43, pp. 207–246.
Enkelmann, V.: Structural Aspects of the Topochemical Polymerization of Diacetylenes. Vol. 63, pp. 91–136.

Ferruti, P. and *Barbucci, R.:* Linear Amino Polymers: Synthesis, Protonation and Complex Formation. Vol. 58, pp. 55–92.
Finkelmann, H. and *Rehage, G.:* Liquid Crystal Side-Chain Polymers. Vol. 60/61, pp. 99–172.
Fischer, H.: Freie Radikale während der Polymerisation, nachgewiesen und identifiziert durch Elektronenspinresonanz. Vol. 5, pp. 463–530.
Flory, P. J.: Molecular Theory of Liquid Crystals. Vol. 59, pp. 1–36.
Ford, W. T. and *Tomoi, M.:* Polymer-Supported Phase Transfer Catalysts Reaction Mechanisms. Vol. 55, pp. 49–104.
Fradet, A. and *Maréchal, E.:* Kinetics and Mechanisms of Polyesterifications. I. Reactions of Diols with Diacids. Vol. 43, pp. 51–144.
Friedrich, K.: Crazes and Shear Bands in Semi-Crystalline Thermoplastics. Vol. 52/53, pp. 225–274
Fujita, H.: Diffusion in Polymer-Diluent Systems. Vol. 3, pp. 1–47.
Funke, W.: Über die Strukturaufklärung vernetzter Makromoleküle, insbesondere vernetzter Polyesterharze, mit chemischen Methoden. Vol. 4, pp. 157–235.

Gal'braikh, L. S. and *Rigovin, Z. A.:* Chemical Transformation of Cellulose. Vol. 14, pp. 87–130.
Galli, G. see Chiellini, E. Vol. 62, pp. 143–170.
Gallot, B. R. M.: Preparation and Study of Block Copolymers with Ordered Structures, Vol. 29, pp. 85–156.
Gandini, A.: The Behaviour of Furan Derivatives in Polymerization Reactions. Vol. 25, pp. 47–96.
Gandini, A. and *Cheradame, H.:* Cationic Polymerization. Initiation with Alkenyl Monomers. Vol. 34/35, pp. 1–289.
Geckeler, K., Pillai, V. N. R., and *Mutter, M.:* Applications of Soluble Polymeric Supports. Vol. 39, pp. 65–94.
Gerrens, H.: Kinetik der Emulsionspolymerisation. Vol. 1, pp. 234–328.
Ghiggino, K. P., Roberts, A. J. and *Phillips, D.:* Time-Resolved Fluorescence Techniques in Polymer and Biopolymer Studies. Vol. 40, pp. 69–167.
Goethals, E. J.: The Formation of Cyclic Oligomers in the Cationic Polymerization of Heterocycles. Vol. 23, pp. 103–130.
Graessley, W. W.: The Etanglement Concept in Polymer Rheology. Vol. 16, pp. 1–179.
Graessley, W. W.: Entagled Linear, Branched and Network Polymer Systems. Molecular Theories. Vol. 47, pp. 67–117.
Grebowicz, J. see Wunderlich, B. Vol. 60/61, pp. 1–60.

Hagihara, N., Sonogashira, K. and *Takahashi, S.:* Linear Polymers Containing Transition Metals in the Main Chain. Vol. 41, pp. 149–179.

Hasegawa, M.: Four-Center Photopolymerization in the Crystalline State. Vol. 42, pp. 1–49.
Hay, A. S.: Aromatic Polyethers. Vol. 4, pp. 496–527.
Hayakawa, R. and *Wada, Y.:* Piezoelectricity and Related Properties of Polymer Films. Vol. 11, pp. 1–55.
Heidemann, E. and *Roth, W.:* Synthesis and Investigation of Collagen Model Peptides. Vol. 43, pp. 145–205.
Heitz, W.: Polymeric Reagents. Polymer Design, Scope, and Limitations. Vol. 23, pp. 1–23.
Helfferich, F.: Ionenaustausch. Vol. 1, pp. 329–381.
Hendra, P. J.: Laser-Raman Spectra of Polymers. Vol. 6, pp. 151–169.
Hendrix, J.: Position Sensitive "X-ray Detectors". Vol. 67, pp. 59–98.
Henrici-Olivé, G. und *Olivé, S.:* Kettenübertragung bei der radikalischen Polymerisation. Vol. 2, pp. 496–577.
Henrici-Olivé, G. und *Olivé, S.:* Koordinative Polymerisation an löslichen Übergangsmetall-Katalysatoren. Vol. 6, pp. 421–472.
Henrici-Olivé, G. and *Olivé, S.:* Oligomerization of Ethylene with Soluble Transition-Metal Catalysts. Vol. 15, pp. 1–30.
Henrici-Olivé, G. and *Olivé, S.:* Molecular Interactions and Macroscopic Properties of Polyacrylonitrile and Model Substances. Vol. 32, pp. 123–152.
Henrici-Olivé, G. and *Olivé, S.:* The Chemistry of Carbon Fiber Formation from Polyacrylonitrile. Vol. 51, pp. 1–60.
Hermans, Jr., J., Lohr, D. and *Ferro, D.:* Treatment of the Folding and Unfolding of Protein Molecules in Solution According to a Lattic Model. Vol. 9, pp. 229–283.
Higashimura, T. and *Sawamoto, M.:* Living Polymerization and Selective Dimerization: Two Extremes of the Polymer Synthesis by Cationic Polymerization. Vol. 62, pp. 49–94.
Hoffman, A. S.: Ionizing Radiation and Gas Plasma (or Glow) Discharge Treatments for Preparation of Novel Polymeric Biomaterials. Vol. 57, pp. 141–157.
Holzmüller, W.: Molecular Mobility, Deformation and Relaxation Processes in Polymers. Vol. 26, pp. 1–62.
Hutchison, J. and *Ledwith, A.:* Photoinitiation of Vinyl Polymerization by Aromatic Carbonyl Compounds. Vol. 14, pp. 49–86.

Iizuka, E.: Properties of Liquid Crystals of Polypeptides: with Stress on the Electromagnetic Orientation. Vol. 20, pp. 79–107.
Ikada, Y.: Characterization of Graft Copolymers. Vol. 29, pp. 47–84.
Ikada, Y.: Blood-Compatible Polymers. Vol. 57, pp. 103–140.
Imanishi, Y.: Synthese, Conformation, and Reactions of Cyclic Peptides. Vol. 20, pp. 1–77.
Inagaki, H.: Polymer Separation and Characterization by Thin-Layer Chromatography. Vol. 24, pp. 189–237.
Inoue, S.: Asymmetric Reactions of Synthetic Polypeptides. Vol. 21, pp. 77–106.
Ise, N.: Polymerizations under an Electric Field. Vol. 6, pp. 347–376.
Ise, N.: The Mean Activity Coefficient of Polyelectrolytes in Aqueous Solutions and Its Related Properties. Vol. 7, pp. 536–593.
Isihara, A.: Intramolecular Statistics of a Flexible Chain Molecule. Vol. 7, pp. 449–476.
Isihara, A.: Irreversible Processes in Solutions of Chain Polymers. Vol. 5, pp. 531–567.
Isihara, A. and *Guth, E.:* Theory of Dilute Macromolecular Solutions. Vol. 5, pp. 233–260.
Iwatsuki, S.: Polymerization of Quinodimethane Compounds. Vol. 58, pp. 93–120.

Janeschitz-Kriegl, H.: Flow Birefrigence of Elastico-Viscous Polymer Systems. Vol. 6, pp. 170–318.
Jenkins, R. and *Porter, R. S.:* Upertubed Dimensions of Stereoregular Polymers. Vol. 36, pp. 1–20.
Jenngins, B. R.: Electro-Optic Methods for Characterizing Macromolecules in Dilute Solution. Vol. 22, pp. 61–81.
Johnston, D. S.: Macrozwitterion Polymerization. Vol. 42, pp. 51–106.

Kamachi, M.: Influence of Solvent on Free Radical Polymerization of Vinyl Compounds. Vol. 38, pp. 55–87.
Kaneko, M. and *Yamada, A.:* Solar Energy Conversion by Functional Polymers. Vol. 55, pp. 1–48.

Kawabata, S. and *Kawai, H.:* Strain Energy Density Functions of Rubber Vulcanizates from Biaxial Extension. Vol. 24, pp. 89–124.
Kennedy, J. P. and *Chou, T.:* Poly(isobutylene-*co*-β-Pinene): A New Sulfur Vulcanizable, Ozone Resistant Elastomer by Cationic Isomerization Copolymerization. Vol. 21, pp. 1–39.
Kennedy, J. P. and *Delvaux, J. M.:* Synthesis, Characterization and Morphology of Poly(butadiene-g-Styrene). Vol. 38, pp. 141–163.
Kennedy, J. P. and *Gillham, J. K.:* Cationic Polymerization of Olefins with Alkylaluminium Initiators. Vol. 10, pp. 1–33.
Kennedy, J. P. and *Johnston, J. E.:* The Cationic Isomerization Polymerization of 3-Methyl-1-butene and 4-Methyl-1-pentene. Vol. 19, pp. 57–95.
Kennedy, J. P. and *Langer, Jr., A. W.:* Recent Advances in Cationic Polymerization. Vol. 3, pp. 508–580.
Kennedy, J. P. and *Otsu, T.:* Polymerization with Isomerization of Monomer Preceding Propagation. Vol. 7, pp. 369–385.
Kennedy, J. P. and *Rengachary, S.:* Correlation Between Cationic Model and Polymerization Reactions of Olefins. Vol. 14, pp. 1–48.
Kennedy, J. P. and *Trivedi, P. D.:* Cationic Olefin Polymerization Using Alkyl Halide — Alkylaluminium Initiator Systems. I. Reactivity Studies. II. Molecular Weight Studies. Vol. 28, pp. 83–151.
Kennedy, J. P., Chang, V. S. C. and *Guyot, A.:* Carbocationic Synthesis and Characterization of Polyolefins with Si–H and Si–Cl Head Groups. Vol. 43, pp. 1–50.
Khoklov, A. R. and *Grosberg, A. Yu.:* Statistical Theory of Polymeric Lyotropic Liquid Crystals. Vol. 41, pp. 53–97.
Kinloch, A. J.: Mechanics and Mechanisms of Fracture of Thermosetting Epoxy Polymers. Vol. 72, pp. 45–68.
Kissin, Yu. V.: Structures of Copolymers of High Olefins. Vol. 15, pp. 91–155.
Kitagawa, T. and *Miyazawa, T.:* Neutron Scattering and Normal Vibrations of Polymers. Vol. 9, pp. 335–414.
Kitamaru, R. and *Horii, F.:* NMR Approach to the Phase Structure of Linear Polyethylene. Vol. 26, pp. 139–180.
Knappe, W.: Wärmeleitung in Polymeren. Vol. 7, pp. 477–535.
Koenig, J. L.: Fourier Transforms Infrared Spectroscopy of Polymers, Vol. 54, pp. 87–154.
Kolařík, J.: Secondary Relaxations in Glassy Polymers: Hydrophilic Polymethacrylates and Polyacrylates: Vol. 46, pp. 119–161.
Koningsveld, R.: Preparative and Analytical Aspects of Polymer Fractionation. Vol. 7.
Kovacs, A. J.: Transition vitreuse dans les polymers amorphes. Etude phénoménologique. Vol. 3, pp. 394–507.
Krässig, H. A.: Graft Co-Polymerization of Cellulose and Its Derivatives. Vol. 4, pp. 111–156.
Kramer, E. J.: Microscopic and Molecular Fundamentals of Crazing. Vol. 52/53, pp. 1–56
Kraus, G.: Reinforcement of Elastomers by Carbon Black. Vol. 8, pp. 155–237.
Kreutz, W. and *Welte, W.:* A General Theory for the Evaluation of X-Ray Diagrams of Biomembranes and Other Lamellar Systems. Vol. 30, pp. 161–225.
Krimm, S.: Infrared Spectra of High Polymers. Vol. 2, pp. 51–72.
Kuhn, W., Ramel, A., Walters, D. H., Ebner, G. and *Kuhn, H. J.:* The Production of Mechanical Energy from Different Forms of Chemical Energy with Homogeneous and Cross-Striated High Polymer Systems. Vol. 1, pp. 540–592.
Kunitake, T. and *Okahata, Y.:* Catalytic Hydrolysis by Synthetic Polymers. Vol. 20, pp. 159–221.
Kurata, M. and *Stockmayer, W. H.:* Intrinsic Viscosities and Unperturbed Dimensions of Long Chain Molecules. Vol. 3, pp. 196–312.

Ledwith, A. and *Sherrington, D. C.:* Stable Organic Cation Salts: Ion Pair Equilibria and Use in Cationic Polymerization. Vol. 19, pp. 1–56.
Ledwith, A. see Chiellini, E. Vol. 62, pp. 143–170.
Lee, C.-D. S. and *Daly, W. H.:* Mercaptan-Containing Polymers. Vol. 15, pp. 61–90.
Lindberg, J. J. and *Hortling, B.:* Cross Polarization — Magic Angle Spinning NMR Studies of Carbohydrates and Aromatic Polymers. Vol. 66, pp. 1–22.

Lipatov, Y. S.: Relaxation and Viscoelastic Properties of Heterogeneous Polymeric Compositions. Vol. 22, pp. 1–59.
Lipatov, Y. S.: The Iso-Free-Volume State and Glass Transitions in Amorphous Polymers: New Development of the Theory. Vol. 26, pp. 63–104.
Luston, J. and *Vašš, F.:* Anionic Copolymerization of Cyclic Ethers with Cyclic Anhydrides. Vol. 56, pp. 91–133.

Madec, J.-P. and *Maréchal, E.:* Kinetics and Mechanisms of Polyesterifications. II. Reactions of Diacids with Diepoxides. Vol. 71, pp. 153–228.
Mano, E. B. and *Coutinho, F. M. B.:* Grafting on Polyamides. Vol. 19, pp. 97–116.
Maréchal, E. see Madec, J.-P. Vol. 71, pp. 153–228.
Mark, J. E.: The Use of Model Polymer Networks to Elucidate Molecular Aspects of Rubberlike Elasticity. Vol. 44, pp. 1–26.
Mark, J. E. see Queslel, J. P. Vol. 71, pp. 229–248.
Maser, F., Bode, K., Pillai, V. N. R. and *Mutter, M.:* Conformational Studies on Model Peptides. Their Contribution to Synthetic, Structural and Functional Innovations on Proteins. Vol. 65, pp. 177–214.
McGrath, J. E. see Yorkgitis, E. M. Vol. 72, pp. 79–110.
McIntyre, J. E. see Dobb, M. G. Vol. 60/61, pp. 61–98.
Meerwall v., E., D.: Self-Diffusion in Polymer Systems. Measured with Field-Gradient Spin Echo NMR Methods, Vol. 54, pp. 1–29.
Mengoli, G.: Feasibility of Polymer Film Coating Through Electroinitiated Polymerization in Aqueous Medium. Vol. 33, pp. 1–31.
Meyerhoff, G.: Die viscosimetrische Molekulargewichtsbestimmung von Polymeren. Vol. 3, pp. 59–105.
Millich, F.: Rigid Rods and the Characterization of Polyisocyanides. Vol. 19, pp. 117–141.
Möller, M.: Cross Polarization — Magic Angle Sample Spinning NMR Studies. With Respect to the Rotational Isomeric States of Saturated Chain Molecules. Vol. 66, pp. 59–80.
Morawetz, H.: Specific Ion Binding by Polyelectrolytes. Vol. 1, pp. 1–34.
Morgan, R. J.: Structure-Property Relations of Epoxies Used as Composite Matrices. Vol. 72, pp. 1–44.
Morin, B. P., Breusova, I. P. and *Rogovin, Z. A.:* Structural and Chemical Modifications of Cellulose by Graft Copolymerization. Vol. 42, pp. 139–166.
Mulvaney, J. E., Oversberger, C. C. and *Schiller, A. M.:* Anionic Polymerization. Vol. 3, pp. 106–138.

Nakase, Y., Kurijama, I. and *Odajima, A.:* Analysis of the Fine Structure of Poly(Oxymethylene) Prepared by Radiation-Induced Polymerization in the Solid State. Vol. 65, pp. 79–134.
Neuse, E.: Aromatic Polybenzimidazoles. Syntheses, Properties, and Applications. Vol. 47, pp. 1–42.
Nicolais, L. see Apicella, A. Vol. 72, pp. 69–78.

Ober, Ch. K., Jin, J.-I. and *Lenz, R. W.:* Liquid Crystal Polymers with Flexible Spacers in the Main Chain. Vol. 59, pp. 103–146.
Okubo, T. and *Ise, N.:* Synthetic Polyelectrolytes as Models of Nucleic Acids and Esterases. Vol. 25, pp. 135–181.
Osaki, K.: Viscoelastic Properties of Dilute Polymer Solutions. Vol. 12, pp. 1–64.
Oster, G. and *Nishijima, Y.:* Fluorescence Methods in Polymer Science. Vol. 3, pp. 313–331.
Otsu, T. see Sato, T. Vol. 71, pp. 41–78.
Overberger, C. G. and *Moore, J. A.:* Ladder Polymers. Vol. 7, pp. 113–150.

Packirisamy, S. see Biswas, M. Vol. 70, pp. 71–118.
Papkov, S. P.: Liquid Crystalline Order in Solutions of Rigid-Chain Polymers. Vol. 59, pp. 75–102.
Patat, F., Killmann, E. und *Schiebener, C.:* Die Absorption von Makromolekülen aus Lösung. Vol. 3, pp. 332–393.

Patterson, G. D.: Photon Correlation Spectroscopy of Bulk Polymers. Vol. 48, pp. 125–159.
Penczek, S., Kubisa, P. and *Matyjaszewski, K.:* Cationic Ring-Opening Polymerization of Heterocyclic Monomers. Vol. 37, pp. 1–149.
Penczek, S., Kubisa, P. and *Matyjaszewski, K.:* Cationic Ring-Opening Polymerization; 2. Synthetic Applications. Vol. 68/69, pp. 1–298.
Peticolas, W. L.: Inelastic Laser Light Scattering from Biological and Synthetic Polymers. Vol. 9, pp. 285–333.
Petropoulos, J. H.: Membranes with Non-Homogeneous Sorption Properties. Vol. 64, pp. 85–134.
Pino, P.: Optically Active Addition Polymers. Vol. 4, pp. 393–456.
Pitha, J.: Physiological Activities of Synthetic Analogs of Polynucleotides. Vol. 50, pp. 1–16.
Platé, N. A. and *Noak, O. V.:* A Theoretical Consideration of the Kinetics and Statistics of Reactions of Functional Groups of Macromolecules. Vol. 31, pp. 133–173.
Platé, N. A. see Shibaev, V. P. Vol. 60/61, pp. 173–252.
Plesch, P. H.: The Propagation Rate-Constants in Cationic Polymerisations. Vol. 8, pp. 137–154.
Porod, G.: Anwendung und Ergebnisse der Röntgenkleinwinkelstreuung in festen Hochpolymeren. Vol. 2, pp. 363–400.
Pospíšil, J.: Transformations of Phenolic Antioxidants and the Role of Their Products in the Long-Term Properties of Polyolefins. Vol. 36, pp. 69–133.
Postelnek, W., Coleman, L. E., and *Lovelace, A. M.:* Fluorine-Containing Polymers. I. Fluorinated Vinyl Polymers with Functional Groups, Condensation Polymers, and Styrene Polymers. Vol. 1, pp. 75–113.

Queslel, J. P. and *Mark, J. E.:* Molecular Interpretation of the Moduli of Elastomeric Polymer Networks of Know Structure. Vol. 65, pp. 135–176.
Queslel, J. P. and *Mark, J. E.:* Swelling Equilibrium Studies of Elastomeric Network Structures. Vol. 71, pp. 229–248.

Rehage, G. see Finkelmann, H. Vol. 60/61, pp. 99–172.
Rempp, P. F. and *Franta, E.:* Macromonomers: Synthesis, Characterization and Applications. Vol. 58, pp. 1–54.
Rempp, P., Herz, J., and *Borchard, W.:* Model Networks. Vol. 26, pp. 107–137.
Richards, R. W.: Small Angle Neutron Scattering from Block Copolymers. Vol. 71, pp. 1–40.
Rigbi, Z.: Reinforcement of Rubber by Carbon Black. Vol. 36, pp. 21–68.
Rogovin, Z. A. and *Gabrielyan, G. A.:* Chemical Modifications of Fibre Forming Polymers and Copolymers of Acrylonitrile. Vol. 25, pp. 97–134.
Roha, M.: Ionic Factors in Steric Control. Vol. 4, pp. 353–392.
Roha, M.: The Chemistry of Coordinate Polymerization of Dienes. Vol. 1, pp. 512–539.
Rostami, S. see Walsh, D. J. Vol. 70, pp. 119–170.

Safford, G. J. and *Naumann, A. W.:* Low Frequency Motions in Polymers as Measured by Neutron Inelastic Scattering. Vol. 5, pp. 1–27.
Sato, T. and *Otsu, T.:* Formation of Living Propagating Radicals in Microspheres and Their Use in the Synthesis of Block Copolymers. Vol. 71, pp. 41–78.
Sauer, J. A. and *Chen, C. C.:* Crazing and Fatigue Behavior in One and Two Phase Glassy Polymers. Vol. 52/53, pp. 169–224
Sawamoto, M. see Higashimura, T. Vol. 62, pp. 49–94.
Schuerch, C.: The Chemical Synthesis and Properties of Polysaccharides of Biomedical Interest. Vol. 10, pp. 173–194.
Schulz, R. C. und *Kaiser, E.:* Synthese und Eigenschaften von optisch aktiven Polymeren. Vol. 4, pp. 236–315.
Seanor, D. A.: Charge Transfer in Polymers. Vol. 4, pp. 317–352.
Semerak, S. N. and *Frank, C. W.:* Photophysics of Excimer Formation in Aryl Vinyl Polymers, Vol. 54, pp. 31–85.
Seidl, J., Malinský, J., Dušek, K. und *Heitz, W.:* Makroporöse Styrol-Divinylbenzol-Copolymere und ihre Verwendung in der Chromatographie und zur Darstellung von Ionenaustauschern. Vol. 5, pp. 113–213.

Semjonow, V.: Schmelzviskositäten hochpolymerer Stoffe. Vol. 5, pp. 387–450.
Semlyen, J. A.: Ring-Chain Equilibria and the Conformations of Polymer Chains. Vol. 21, pp. 41–75.
Sharkey, W. H.: Polymerizations Through the Carbon-Sulphur Double Bond. Vol. 17, pp. 73–103.
Shibaev, V. P. and *Platé, N. A.:* Thermotropic Liquid-Crystalline Polymers with Mesogenic Side Groups. Vol. 60/61, pp. 173–252.
Shimidzu, T.: Cooperative Actions in the Nucleophile-Containing Polymers. Vol. 23, pp. 55–102.
Shutov, F. A.: Foamed Polymers Based on Reactive Oligomers, Vol. 39, pp. 1–64.
Shutov, F. A.: Foamed Polymers. Cellular Structure and Properties. Vol. 51, pp. 155–218.
Siesler, H. W.: Rheo-Optical Fourier-Transform Infrared Spectroscopy: Vibrational Spectra and Mechanical Properties of Polymers. Vol. 65, pp. 1–78.
Silvestri, G., Gambino, S., and *Filardo, G.:* Electrochemical Production of Initiators for Polymerization Processes. Vol. 38, pp. 27–54.
Sixl, H.: Spectroscopy of the Intermediate States of the Solid State Polymerization Reaction in Diacetylene Crystals. Vol. 63, pp. 49–90.
Slichter, W. P.: The Study of High Polymers by Nuclear Magnetic Resonance. Vol. 1, pp. 35–74.
Small, P. A.: Long-Chain Branching in Polymers. Vol. 18.
Smets, G.: Block and Graft Copolymers. Vol. 2, pp. 173–220.
Smets, G.: Photochromic Phenomena in the Solid Phase. Vol. 50, pp. 17–44.
Sohma, J. and *Sakaguchi, M.:* ESR Studies on Polymer Radicals Produced by Mechanical Destruction and Their Reactivity. Vol. 20, pp. 109–158.
Solaro, R. see Chiellini, E. Vol. 62, pp. 143–170.
Sotobayashi, H. und *Springer, J.:* Oligomere in verdünnten Lösungen. Vol. 6, pp. 473–548.
Sperati, C. A. and *Starkweather, Jr., H. W.:* Fluorine-Containing Polymers. II. Polytetrafluoroethylene. Vol. 2, pp. 465–495.
Spiess, H. W.: Deuteron NMR — A new Tool for Studying Chain Mobility and Orientation in Polymers. Vol. 66, pp. 23–58.
Sprung, M. M.: Recent Progress in Silicone Chemistry. I. Hydrolysis of Reactive Silane Intermediates, Vol. 2, pp. 442–464.
Stahl, E. and *Brüderle, V.:* Polymer Analysis by Thermofractography. Vol. 30, pp. 1–88.
Stannett, V. T., Koros, W. J., Paul, D. R., Lonsdale, H. K., and *Baker, R. W.:* Recent Advances in Membrane Science and Technology. Vol. 32, pp. 69–121.
Staverman, A. J.: Properties of Phantom Networks and Real Networks. Vol. 44, pp. 73–102.
Stauffer, D., Coniglio, A. and *Adam, M.:* Gelation and Critical Phenomena. Vol. 44, pp. 103–158.
Stille, J. K.: Diels-Alder Polymerization. Vol. 3, pp. 48–58.
Stolka, M. and *Pai, D.:* Polymers with Photoconductive Properties. Vol. 29, pp. 1–45.
Stuhrmann, H.: Resonance Scattering in Macromolecular Structure Research. Vol. 67, pp. 123–164.
Subramanian, R. V.: Electroinitiated Polymerization on Electrodes. Vol. 33, pp. 35–58.
Sumitomo, H. and *Hashimoto, K.:* Polyamides as Barrier Materials. Vol. 64, pp. 55–84.
Sumitomo, H. and *Okada, M.:* Ring-Opening Polymerization of Bicyclic Acetals, Oxalactone, and Oxalactam. Vol. 28, pp. 47–82.
Szegö, L.: Modified Polyethylene Terephthalate Fibers. Vol. 31, pp. 89–131.
Szwarc, M.: Termination of Anionic Polymerization. Vol. 2, pp. 275–306.
Szwarc, M.: The Kinetics and Mechanism of N-carboxy-α-amino-acid Anhydride (NCA) Polymerization to Poly-amino Acids. Vol. 4, pp. 1–65.
Szwarc, M.: Thermodynamics of Polymerization with Special Emphasis on Living Polymers. Vol. 4, pp. 457–495.
Szwarc, M.: Living Polymers and Mechanisms of Anionic Polymerization. Vol. 49, pp. 1–175.

Takahashi, A. and *Kawaguchi, M.:* The Structure of Macromolecules Adsorbed on Interfaces. Vol. 46, pp. 1–65.
Takemoto, K. and *Inaki, Y.:* Synthetic Nucleic Acid Analogs. Preparation and Interactions. Vol. 41, pp. 1–51.
Tani, H.: Stereospecific Polymerization of Aldehydes and Epoxides. Vol. 11, pp. 57–110.
Tate, B. E.: Polymerization of Itaconic Acid and Derivatives. Vol. 5, pp. 214–232.
Tazuke, S.: Photosensitized Charge Transfer Polymerization. Vol. 6, pp. 321–346.
Teramoto, A. and *Fujita, H.:* Conformation-dependent Properties of Synthetic Polypeptides in the Helix-Coil Transition Region. Vol. 18, pp. 65–149.

Theocaris, P. S.: The Mesophase and its Influence on the Mechanical Behavior of Composites. Vol. 66, pp. 149–188.
Thomas, W. M.: Mechanismus of Acrylonitrile Polymerization. Vol. 2, pp. 401–441.
Tieke, B.: Polymerization of Butadiene and Butadiyne (Diacetylene) Derivatives in Layer Structures. Vol. 71, pp. 79–152.
Tobolsky, A. V. and *DuPré, D. B.:* Macromolecular Relaxation in the Damped Torsional Oscillator and Statistical Segment Models. Vol. 6, pp. 103–127.
Tosi, C. and *Ciampelli, F.:* Applications of Infrared Spectroscopy to Ethylene-Propylene Copolymers. Vol. 12, pp. 87–130.
Tosi, C.: Sequence Distribution in Copolymers: Numerical Tables. Vol. 5, pp. 451–462.
Tran, C. see Yorkgitis, E. M. Vol. 72, pp. 79–110.
Tsuchida, E. and *Nishide, H.:* Polymer-Metal Complexes and Their Catalytic Activity. Vol. 24, pp. 1–87.
Tsuji, K.: ESR Study of Photodegradation of Polymers. Vol. 12, pp. 131–190.
Tsvetkov, V. and *Andreeva, L.:* Flow and Electric Birefringence in Rigid-Chain Polymer Solutions. Vol. 39, pp. 95–207.
Tuzar, Z., Kratochvil, P., and *Bohdanecký, M.:* Dilute Solution Properties of Aliphatic Polyamides. Vol. 30, pp. 117–159.

Uematsu, I. and *Uematsu, Y.:* Polypeptide Liquid Crystals. Vol. 59, pp. 37–74.

Valvassori, A. and *Sartori, G.:* Present Status of the Multicomponent Copolymerization Theory. Vol. 5, pp. 28–58.
Viovy, J. L. and *Monnerie, L.:* Fluorescence Anisotropy Technique Using Synchrotron Radiation as a Powerful Means for Studying the Orientation Correlation Functions of Polymer Chains. Vol. 67, pp. 99–122.
Voigt-Martin, I.: Use of Transmission Electron Microscopy to Obtain Quantitative Information About Polymers. Vol. 67, pp. 195–218.
Voorn, M. J.: Phase Separation in Polymer Solutions. Vol. 1, pp. 192–233.

Walsh, D. J., Rostami, S.: The Miscibility of High Polymers: The Role of Specific Interactions. Vol. 70, pp. 119–170.
Ward, I. M.: Determination of Molecular Orientation by Spectroscopic Techniques. Vol. 66, pp. 81–116.
Ward, I.M.: The Preparation, Structure and Properties of Ultra-High Modulus Flexible Polymers. Vol. 70, pp. 1–70.
Werber, F. X.: Polymerization of Olefins on Supported Catalysts. Vol. 1, pp. 180–191.
Wichterle, O., Šebenda, J., and *Králíček, J.:* The Anionic Polymerization of Caprolactam. Vol. 2, pp. 578–595.
Wilkes, G. L.: The Measurement of Molecular Orientation in Polymeric Solids. Vol. 8, pp. 91–136.
Wilkes, G. L. see Yorkgitis, E. M. Vol. 72, pp. 79–110.
Williams, G.: Molecular Aspects of Multiple Dielectric Relaxation Processes in Solid Polymers. Vol. 33, pp. 59–92.
Williams, J. G.: Applications of Linear Fracture Mechanics. Vol. 27, pp. 67–120.
Wöhrle, D.: Polymere aus Nitrilen. Vol. 10, pp. 35–107.
Wöhrle, D.: Polymer Square Planar Metal Chelates for Science and Industry. Synthesis, Properties and Applications. Vol. 50, pp. 45–134.
Wolf, B. A.: Zur Thermodynamik der enthalpisch und der entropisch bedingten Entmischung von Polymerlösungen. Vol. 10, pp. 109–171.
Woodward, A. E. and *Sauer, J. A.:* The Dynamic Mechanical Properties of High Polymers at Low Temperatures. Vol. 1, pp. 114–158.
Wunderlich, B.: Crystallization During Polymerization. Vol. 5, pp. 568–619.
Wunderlich, B. and *Baur, H.:* Heat Capacities of Linear High Polymers. Vol. 7, pp. 151–368.

Wunderlich, B. and *Grebowicz, J.:* Thermotropic Mesophases and Mesophase Transitions of Linear, Flexible Macromolecules. Vol. 60/61, pp. 1–60.
Wrasidlo, W.: Thermal Analysis of Polymers. Vol. 13, pp. 1–99.

Yamashita, Y.: Random and Black Copolymers by Ring-Opening Polymerization. Vol. 28, pp. 1–46.
Yamazaki, N.: Electrolytically Initiated Polymerization. Vol. 6, pp. 377–400.
Yamazaki, N. and *Higashi, F.:* New Condensation Polymerizations by Means of Phosphorus Compounds. Vol. 38, pp. 1–25.
Yokoyama, Y. and *Hall, H. K.:* Ring-Opening Polymerization of Atom-Bridged and Bond-Bridged Bicyclic Ethers, Acetals and Orthoesters. Vol. 42, pp. 107–138.
Yorkgitis, E. M., Eiss, N. S. Jr., Tran, C., Wilkes, G. L. and *McGrath, J. E.:* Siloxane-Modified Epoxy Resins. Vol. 72, pp. 79–110.
Yoshida, H. and *Hayashi, K.:* Initiation Process of Radiation-induced Ionic Polymerization as Studied by Electron Spin Resonance. Vol. 6, pp. 401–420.
Young, R. N., Quirk, R. P. and *Fetters, L. J.:* Anionic Polymerizations of Non-Polar Monomers Involving Lithium. Vol. 56, pp. 1–90.
Yuki, H. and *Hatada, K.:* Stereospecific Polymerization of Alpha-Substituted Acrylic Acid Esters. Vol. 31, pp. 1–45.

Zachmann, H. G.: Das Kristallisations- und Schmelzverhalten hochpolymerer Stoffe. Vol. 3, pp. 581–687.
Zaikov, G. E. see *Aseeva, R. M.* Vol. 70, pp. 171–230.
Zakharov, V. A., Bukatov, G. D., and *Yermakov, Y. I.:* On the Mechanism of Olifin Polymerization by Ziegler-Natta Catalysts. Vol. 51, pp. 61–100.
Zambelli, A. and *Tosi, C.:* Stereochemistry of Propylene Polymerization. Vol. 15, pp. 31–60.
Zucchini, U. and *Cecchin, G.:* Control of Molecular-Weight Distribution in Polyolefins Synthesized with Ziegler-Natta Catalytic Systems. Vol. 51, pp. 101–154.

Subject Index

Activation energy 117, 118
Adhesive joint specimens 65
Adhesive, structural 66
Amine addition 123
Amine cure 4, 131–143
Amine-epoxide addition 146
Anhydride 122
Anhydride cure 126, 128–130
Anomalous plasticization 3
ATBN copolymers 2
Autocatalysis 121, 123, 124, 138

Base catalysis 124
Baseline 116, 117
BF_3:amine catalysts 8–18
BF_4^- salt formation 12
Birefringence studies 36
Bisphenol-A-diglycidyl ether epoxies, amine-cured 4
Boron trifluoride complexes 124, 140

Carbonyl IR band 19
Carboxylic acid anhydride 120
Catalysis 121, 140
Catalyst composition in prepregs 12
Cavitation 58
C-fiber epoxy composites 5
— —, prepregs, commercial 7
Chain scission, stress-induced 38
Clusters 3
Cooling, rate of 31
Compact tension geometry 6
Composite matrices, tough 5
Composites 3
—, filament-wound 3
—, hybrid-particulate 52
Composition, effects 51
Computer modeling 33
Copolymers 2
Crack growth, types 49
— initiation and arrest 52
— of tip radius 63
— opening displacement 60
— pinning 59

Crack-tip blunting 60
— micromechanisms 57
Crazing 35, 57
— agents 2
Critical distance 64
— stress 64
Crosslinking 114
— reactions 141
Cross-link density 135
Cross-links, intermolecularly formed 29
CTBN copolymers 2
Cure mechanism 137
— reactions 4, 18, 120
— reactions, chemistry of 28
— —, rates of 28

DDS recrystallization 30
Deformation and failure modes 35
Degradation reactions 20
Diaminodiphenyl sulfone-cured tetraglycidyl-4,4'-diaminodiphenylmethane epoxies 6
Dicyanodiamide 124, 125, 144–149
Differential scanning calorimetry 15
Diffusion control 124, 132, 135
Dilution process 6
Dissolved water 7
Diuron 145, 147
DSC instruments 112–114
DSC peaks, characterization of 17
DTA 113
Dynamic mechanical measurements 32
— — properties 1
— — tests 8

Electron paramagnetic resonance 39
Epoxide consumption, rate of 19
— homopolymerization 19
— impurities 7
— -hydroxyl reactions 21
— IR band 19
— isomerization 21
Epoxides, ether formation from 27
Epoxies, physical structure 31

Epoxy dehydration 26
— modulus 38
— network embrittlement 39
— — structure 32
Epoxy polymer, rubber-modified 53
Esterification 122
Ether crosslinks 28
— formation 27
— IR band 19
Etherification 122
Exothermic reaction 115
Exponential integral 119

Facture topographies 35
Failure criterion 63
Fiberite 934 13
Filament winding epoxies 5
— -wound composites 3
Flory-Huggins 4
Fracture criteria 48
— energy 51
— mechanics 47
— — linear-elastic 49
— toughness 6, 48
Free volume 31
— — changes, moisture-induced 40
— — elements 4
Friction and wear 1
Fourier transform IR spectroscopy 18

Gelation 2, 122, 124
— time 146
Gel particles 32
Glass transition 7, 114, 135, 136, 141
— — temperature 129, 132

Heat flow 115
— flux 112, 113
Heating rate 118
Hercules 3501 13
Homopolymerization 4
Hybrid-particulate composites 52
Hydrogen bonding 5, 6
— — mechanism 8
Hydrophilio sites 4, 6, 8
Hydroxyl catalysis 123, 133, 138
— -epoxide addition 146
— IR band 19

Imidazoles 125, 126, 148, 150
Impact strength 2
Isothermal data 116

Kinetic equations 118–120, 134, 136, 139
Kinetics 117

Langmuir 5

Macroscopic inhomogeneities 31
Mechanical properties 1, 38
Microcavitation 7
Microscopic failure processes 36
Microstructure, multiphase 52
Microvoiding 3, 5
Microvoids 31
Moisture, sorbed 39
Molecular modeling 30
Monoboroester 14
Monuron 145
Morphology 1
Multiphase microstructure 52

Narmo 5208 13
Near infrared spectroscopy 4
Network basic rings 34
— defects 33
— extensibility 33
— morphology 32
— segment extensibility 33
NMR spectroscopy 7
— —, ^{1}H 8
— —, ^{19}F 10
Nodular structure 2
Nucleophic addition 4

Oxidation 22

Particle/matrix adhesion 60
Plastic deformation, homogeneous 35
— flow 35
Plasticization 5, 6, 7
Plasticizer 2
Polimer-diluent interactions 3
Polyethertriamine curing agent 4
Power compensation 112
Prepreg processing viscosity 30
Primary amine — epoxide reaction 23
Propenal formation 26

Rate constant 117
Resins 47
Rings, intramolecularly formed 29
Rubber toughening 2

Scanning electron microscopy 1
Secondary amine — epoxide reaction 23
Service environment aging 39
Shear banding 35
— bands 59
— yielding 57
Shift factor 55
Siloxane-modified epoxy networks 1
Small-angle X-ray scattering 33
Stationary state 121
Stress-FTIR 39
— -intensity factor 48

Subject Index

— theoretical 47
Superposition, time-temperature 55
Swelling stresses 5

Tan σ 8
Temperature scan 116
Tensile stength 38
· TGDDM-DDS epoxy cure reactions 22
— epoxy system, degradation 26
Thermal expansion coefficient 6
— spike 39

Three-point bend geometry 6
Toughening 2
Transmission electron microscopy 33

Viscoelastic yield behaviour 61
Viterfication 2

Water diffusion 3

Yield stresses 36